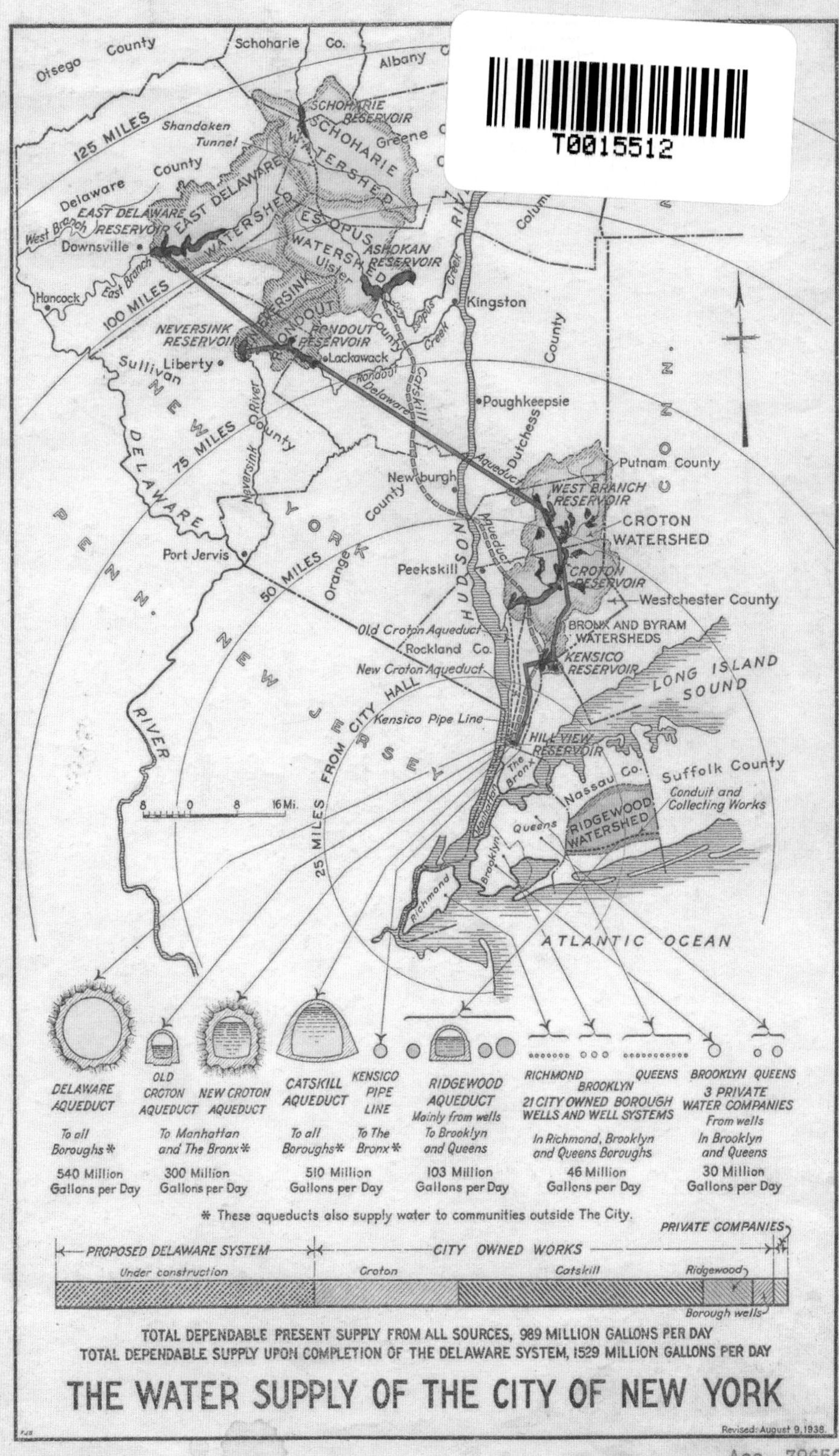

Otsego County
Schoharie Co.
Albany
SCHOHARIE RESERVOIR
SCHOHARIE WATERSHED
Greene
125 MILES
Shandaken Tunnel
Delaware County
EAST DELAWARE RESERVOIR
EAST DELAWARE WATERSHED
West Branch
Downsville
ESOPUS WATERSHED
ASHOKAN RESERVOIR
Ulster
Hancock
East Branch
100 MILES
NEVERSINK
RONDOUT
NEVERSINK RESERVOIR
RONDOUT RESERVOIR
Kingston
Esopus Creek
Lackawack
Liberty
Sullivan County
Rondout
Delaware
Catskill
Aqueduct
Poughkeepsie
Dutchess County
NEW YORK
75 MILES
DELAWARE
Neversink River
Putnam County
Newburgh
WEST BRANCH RESERVOIR
CROTON WATERSHED
Orange County
PENN.
Port Jervis
50 MILES
Peekskill
HUDSON
CROTON RESERVOIR
Westchester County
CONN.
Old Croton Aqueduct
Rockland Co.
BRONX AND BYRAM WATERSHEDS
KENSICO RESERVOIR
New Croton Aqueduct
LONG ISLAND SOUND
NEW JERSEY
25 MILES FROM CITY HALL
Kensico Pipe Line
HILL VIEW RESERVOIR
RIVER
The Bronx
Nassau Co.
Suffolk County
Conduit and Collecting Works
8 0 8 16 Mi.
Manhattan
Queens
RIDGEWOOD WATERSHED
Brooklyn
Richmond
ATLANTIC OCEAN
DELAWARE AQUEDUCT
To all Boroughs*
540 Million Gallons per Day
OLD CROTON AQUEDUCT
NEW CROTON AQUEDUCT
To Manhattan and The Bronx*
300 Million Gallons per Day
CATSKILL AQUEDUCT
To all Boroughs*
KENSICO PIPE LINE
To The Bronx*
510 Million Gallons per Day
RIDGEWOOD AQUEDUCT
Mainly from wells
To Brooklyn and Queens
103 Million Gallons per Day
RICHMOND
BROOKLYN
QUEENS
21 CITY OWNED BOROUGH WELLS AND WELL SYSTEMS
In Richmond, Brooklyn and Queens Boroughs
46 Million Gallons per Day
BROOKLYN
QUEENS
3 PRIVATE WATER COMPANIES
From wells
In Brooklyn and Queens
30 Million Gallons per Day
* These aqueducts also supply water to communities outside The City.
PROPOSED DELAWARE SYSTEM
CITY OWNED WORKS
PRIVATE COMPANIES
Under construction
Croton
Catskill
Ridgewood
Borough wells
TOTAL DEPENDABLE PRESENT SUPPLY FROM ALL SOURCES, 989 MILLION GALLONS PER DAY
TOTAL DEPENDABLE SUPPLY UPON COMPLETION OF THE DELAWARE SYSTEM, 1529 MILLION GALLONS PER DAY
THE WATER SUPPLY OF THE CITY OF NEW YORK
Revised: August 9, 1938.

DOWNSVILLE - N.Y
FLOOD AUG 24-1933.

NINETEEN
RESERVOIRS

Also by Lucy Sante

I Heard Her Call My Name

Maybe the People Would Be the Times

The Other Paris

Folk Photography

Kill All Your Darlings

The Factory of Facts

Evidence

Low Life

NINETEEN RESERVOIRS

On Their Creation *and the* Promise of Water *for* New York City

LUCY SANTE

With photographs by Tim Davis

THE EXPERIMENT
NEW YORK

Nineteen Reservoirs: *On Their Creation and the Promise of Water for New York City*

Pages 188–89 are a continuation of this copyright page.

An earlier version of this book was published as "Reservoir: Nature, Culture, Infrastructure" by *Places Journal* in 2020. First published in revised form by The Experiment, LLC, in 2022. This paperback edition first published in 2024.

The Experiment, LLC
220 East 23rd Street, Suite 600
New York, NY 10010-4658
theexperimentpublishing.com

This work was supported by a grant from the Graham Foundation for Advanced Studies in the Fine Arts.

Graham Foundation

Library of Congress Cataloging-in-Publication Data

Names: Sante, Lucy, author.
Title: Nineteen reservoirs : on their creation and the promise of water for New York City / Lucy Sante.
Description: New York : The Experiment, 2022. | Includes bibliographical references and index.
Identifiers: LCCN 2022007559 (print) | LCCN 2022007560 (ebook) | ISBN 9781615198658 | ISBN 9781615198665 (ebook)
Subjects: LCSH: Water-supply--New York (State)--New York--History. | Reservoirs--New York (State)--History.
Classification: LCC TD225.N5 S26 2022 (print) | LCC TD225.N5 (ebook) | DDC 363.6/1097471--dc23/eng/20220608
LC record available at https://lccn.loc.gov/2022007559
LC ebook record available at https://lccn.loc.gov/2022007560

ISBN 978-1-891011-72-6
Ebook ISBN 978-1-61519-866-5

Book and cover design by Beth Bugler
Cover images by Tim Davis; Flickr/NYC Water courtesy of the NYC DEP; Adobe Stock
Author photograph by Bob Krasner

Page i: The New York City water supply system, 1938; pages ii–iii: Downsville flood, postcard, 1933; pages iv–v: The Empire State Building in New York City

Manufactured in China

First paperback printing May 2024
10 9 8 7 6 5 4 3 2 1

Surveyors at Bonticou Crag determining the future path of the Catskill Aqueduct, November 1906

High-pressure hydrant test at Union Square, New York City, 1908

CONTENTS

Prospective map showing a bird's-eye view of points of interest in New York City, 1892

NEW YORK.

, 115 NASSAU ST.

The city and harbor of
New York, 1896

Introduction

New York City became one of the world's great cities in large part because it is one of the world's great natural harbors. From Manhattan Island to Staten Island to Long Island (read: Brooklyn and Queens), the place is surrounded and permeated by water. But none of it is potable—not the East or Harlem Rivers, which are effectively arms of the sea, and not the Hudson River, which in its lower reaches also becomes a tidal estuary, combining sea and fresh water in a varyingly brackish mix for at least the lower half of its 153-mile course, from its mouth in New York Harbor back up to the Federal Dam in Troy near the center of the state. For drinking, cooking, and washing purposes, the first European settlers were able to tap the islands' ponds, streams, and springs. But an ever-expanding population eventually drained such resources—those not already tainted by pollution or disease. And the population galloped on relentlessly: From a thousand or so Dutch nationals in 1650, fourteen years before the British invasion and seventeen before the first wells were dug, the number had jumped to nearly 25,000 by the time of the American Revolution. By 1800, it was 60,000. It was then that the city's political and financial powers recognized a crisis and understood that water would have to be brought in from outside the city.

It was the first of many such realizations and interventions over the succeeding century and a half. Once the idea of importing water had seated itself in the public mind, it was hard to dislodge, for without more reliable sources, the city could not survive. The moral justification for the seizure of water belonging to other regions—the landowners were always compensated, but under the terms of eminent domain they were not permitted to refuse—was invariably that the needs of the many overrode the rights of the few. That was true as far as it went: More and better water would benefit immigrants, freed slaves, disabled veterans, the chronically ill, and the destitute in addition to the middle and upper classes. But although such arguments

Mulberry Street, New York City, circa 1900

were politically convenient, the architects of New York's water-importation schemes across the decades seem to have been rather more concerned with the needs of business. The requirements of the great mass of people reached their ears only on occasions when the city was in the grip of what was called a "water famine," when pipes lacked sufficient pressure to serve the upper floors of buildings, making the whole metropolis a fire hazard. It follows that politicians

Street scene in the Ashokan district near Brown's Station, postcard, circa 1907

and their commissioners tended to think of the extramural regions from which they proposed to pump as essentially uninhabited, since no industry was occurring there. Their attitude, in a word, was colonial.

The people whose land was taken reacted with disbelief, sorrow, anger. That land might have been in their families for generations, might have been the family's sole support, might have been the only home they'd ever known. The city decreed and mobilized and condemned properties for seizure without asking residents' permission, found all sorts of legal subterfuges for denying the value of their fields and homesteads as established by expert

witnesses, lowballed every estimate, treated them with distant contempt. Since the politicians and commissioners were not running for office upstate, they felt no need to package their enterprise as a humanitarian mission; they spoke in numbers and legal precedents. There was already a long history of mistrust between the city and its rural neighbors, who felt themselves sidelined in the political discourse that traveled between Manhattan and the state capitol in Albany: The small farmers and small-town business owners to the north of the great city were routinely caricatured in the press as "apple knockers" and "rubes," were exploited by prosperous urbanites who built summer homes in choice locations and then prosecuted trespassers. That these same remote and implacable beings were now proposing to drown pastures, raze villages, usurp water, and even decree how remaining land should be worked was a shock if not exactly a surprise.

You could say that the very idea of diverting water from somewhere else to benefit the inhabitants of New York City represents a version of the trolley problem. In that ethical thought experiment, you are standing by the switch as a railcar comes barreling along, headed for five people tied up on the tracks. You can pull the lever and turn the car onto a branch line, but there you see one person tied up. Do you do nothing and permit five deaths, or act and cause one? Do you do nothing and impose a water famine on a teeming city, or do you pull the lever and shift the onus onto much more sparsely populated rural areas? There is no satisfactory solution to this dilemma, and New York City had no real choice.

Throughout the nineteenth century, as multiple plans were suggested for bringing in water, those that avoided imposing on the countryside tended to lie on the farther shore of practicality. Unrealized schemes—all of them soberly submitted and considered—ranged from an 1834 proposal to dam the Hudson at Christopher Street in Greenwich Village to one in 1950 for damming it at Haverstraw in Rockland County 40 miles north, a 1905 plan for a pipeline from the Great Lakes, and a 1966 scheme to dam Long Island Sound and turn it into a freshwater lake. The more realistic ideas all involved tapping upstate watercourses, although these raised their own momentous engineering challenges, not least of which was the basic question of how to convey water to the

Looking toward Granton from Apex, NY, postcard, circa 1910

city through many miles of diverse topography and shifting geological properties, let alone the problem of getting it across the broad and deep Hudson.

The system that was eventually built comprises six great reservoirs—Ashokan, Gilboa, Rondout, Neversink, Pepacton, and Cannonsville—that were put in place between 1907 and 1967 in the Catskill/Delaware watershed on the west side of the Hudson. They, along with the older and much smaller Croton system of reservoirs on the east side of the river (not all of which still operate), continue collectively to supply the city with more than 1.1 billion gallons of fresh water every day. The reservoir system has been a great success, even if it required frequent expansion over many years. A half century after its completion, contingency schemes are still intermittently discussed. But aside from the occasional year of drought, no real crisis has appeared to make the New York City Department of Environmental Protection consider immediate plans for expansion beyond the nineteen reservoirs in its system.

Nevertheless, from an upstate perspective, the reservoir system represents at best an imposition and at worst an imperial pillage of the landscape. Twenty-six villages and countless farms, orchards, quarries, and the like were bought for a fraction of their value, demolished, and then submerged, some of them within living memory, leaving broken hearts and fractured communities. The system has further affected a political polarization between upstate and down, city and country, that was already well underway before the first shovel of soil was

Main Street, Union Grove, NY, postcard, circa 1910

removed, and which appears as a microcosm of the urban/rural polarity that continues to unbalance the nation as a whole. The regional ecosystem was altered, including in ways we cannot fully appreciate; no biologists were on hand in 1907 to count the number of endangered species or trace the deleterious effects of lost habitats. My purpose here is not to condemn the reservoir system, without which New York City might have faded into insignificance over the course of the twentieth century, not only squelching its vast financial powers but aborting its function as shelter for millions of people displaced from elsewhere. I would simply like to give an account of the human costs, an overview of the trade-offs, a summary of unintended consequences.

The New York City water supply system, 1917

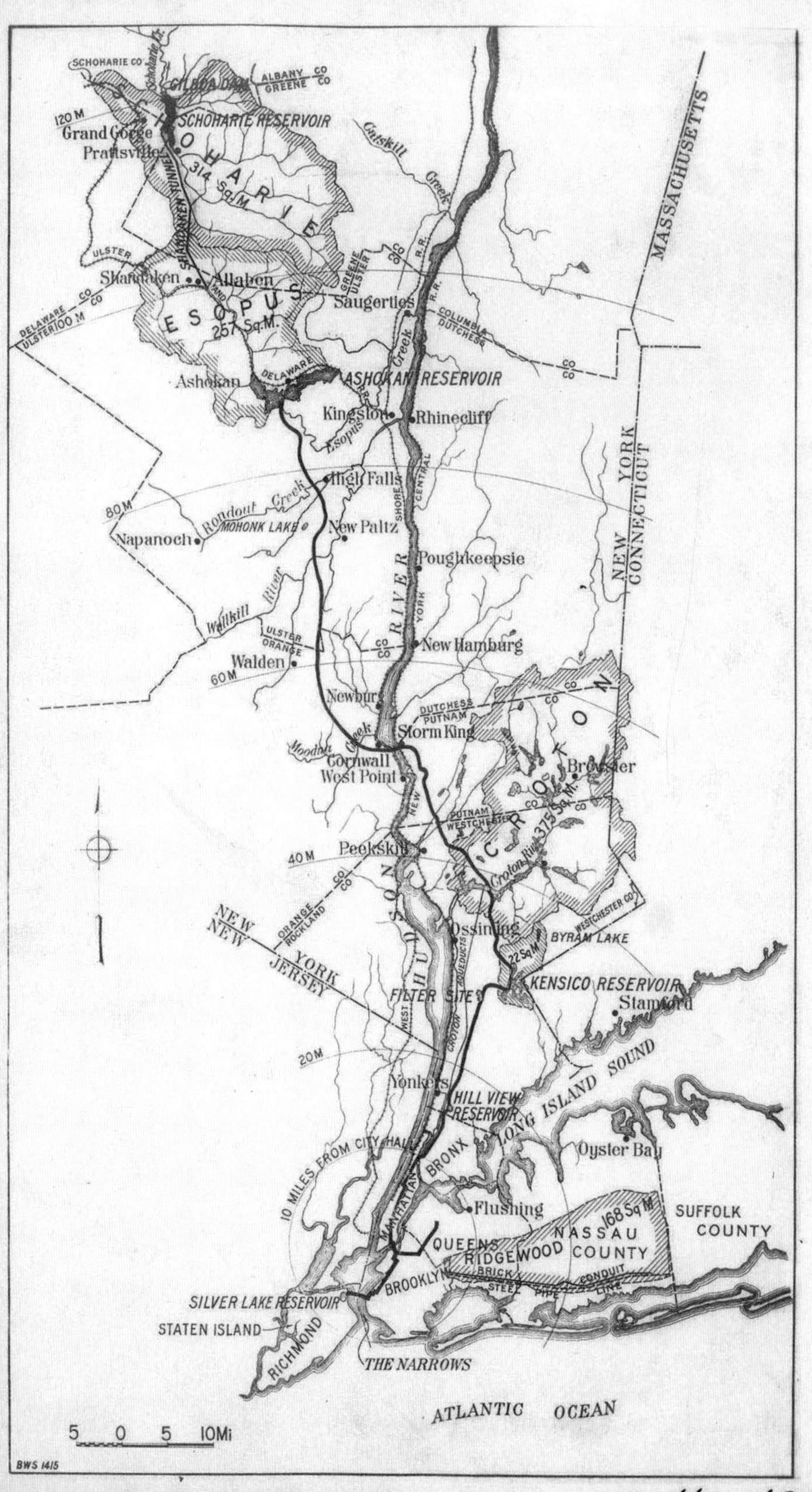
SCHOHARIE CO
Schoharie Cr.
GILBOA DAM
ALBANY CO
GREENE CO
120 M
SCHOHARIE RESERVOIR
Grand Gorge
Prattsville
SCHOHARIE
314 Sq.M.
SHANDAKEN TUNNEL
Catskill
Creek
MASSACHUSETTS
ULSTER
Shandaken
Allaben
ESOPUS
257 Sq.M.
GREENE
ULSTER
CO
R.R.
Saugerties
COLUMBIA
DUTCHESS
DELAWARE CO
ULSTER CO
100 M
Creek
Ashokan
DELAWARE
ASHOKAN RESERVOIR
Kingston
Rhinecliff
Esopus
High Falls
SHORE
CENTRAL
80 M
Rondout
Creek
MOHONK LAKE
New Paltz
Napanoch
Poughkeepsie
NEW YORK
CONNECTICUT
RIVER
YORK
Wallkill
River
ULSTER
ORANGE
CO
New Hamburg
Walden
60 M
Newburg
DUTCHESS
PUTNAM
CO
CROTON
Storm King
Moodna
Creek
Cornwall
West Point
Brewster
NEW
PUTNAM
WESTCHESTER
375 Sq.M.
Peekskill
40 M
Croton River
ORANGE
ROCKLAND
CO
WESTCHESTER CO
NEW YORK
NEW JERSEY
HUDSON
Ossining
AQUEDUCTS
BYRAM LAKE
22 Sq.M.
KENSICO RESERVOIR
FILTER SITE
Stamford
WEST
CROTON
20 M
LONG ISLAND SOUND
Yonkers
HILL VIEW
RESERVOIR
BRONX
Oyster Bay
10 MILES FROM CITY HALL
MANHATTAN
Flushing
168 Sq.M.
SUFFOLK
COUNTY
QUEENS
NASSAU
RIDGEWOOD
COUNTY
BRICK
CONDUIT
BROOKLYN
STEEL PIPE LINE
SILVER LAKE RESERVOIR
STATEN ISLAND
RICHMOND
THE NARROWS
ATLANTIC OCEAN
5 0 5 10 Mi
BWS 1415

Hdq.1606

ALBANY
254 NO. AQUEDUCT
254 NO. AQUEDUCT

1 The Croton System

Panoramic view of the Hudson River during construction of the Catskill Aqueduct, 1907

The yellow fever epidemic of 1798 was the proximate cause for the chartering, a year later, of the Manhattan Company. Although yellow fever was caused by mosquitoes, people of the time attributed its effects—and those of cholera and typhoid fever, which were also rampant—to foul water from various ponds and wells. Accordingly, the company built a 550,000-gallon reservoir on Chambers Street east of Broadway, near what is now Foley Square. The facility, the first in the city's history, served two thousand households—naturally the wealthiest—with 25 miles of mains constructed from hollowed-out tree trunks, which were disinterred by construction several times across the better part of two centuries. (The original plan had involved bringing water to Manhattan from Rye Pond in Westchester County, some 30 miles to the east, but that part of the proposal was never implemented.) The head of the Manhattan Company was Aaron Burr, who had his own scheme

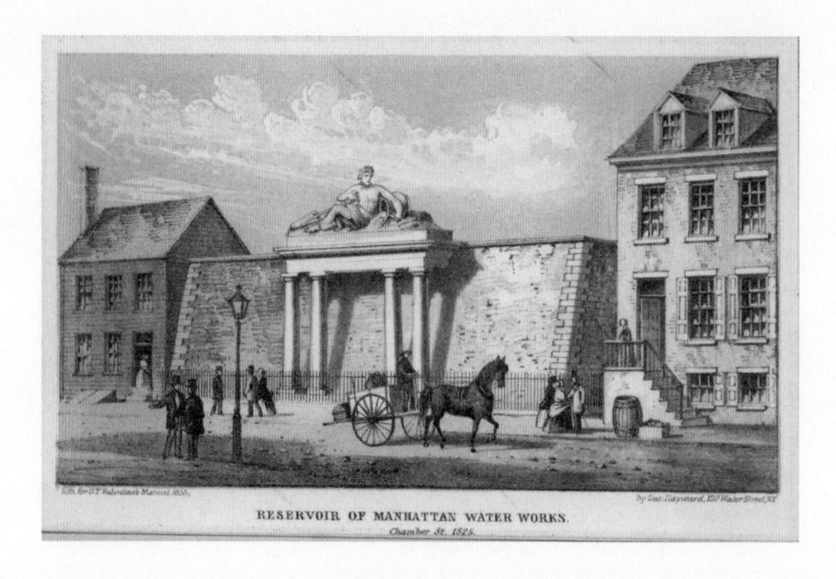

Reservoir of Manhattan waterworks, Chambers Street, 1825

in mind, to which the water-bearing endeavor was subordinate: He introduced into the company's incorporation papers a clause permitting its surplus capital to be used in any transactions not inconsistent with state laws. That allowed for creation of a bank, with which Burr aimed to challenge the Bank of New York, headed by his rival Alexander Hamilton and at that point the city's only chartered bank; the bankers' rivalry was perhaps an underlying cause of the duel in 1804 in which Burr killed Hamilton. (The Bank of New York merged with the Mellon Financial Corporation in 2007; Burr's firm is still in business as Chase Bank.)

The inadequacy of the Chambers Street reservoir was dramatically illustrated by devastating fires in 1828 and 1835, the latter of which destroyed 700 buildings in a seventeen-block area, and by the cholera epidemic of 1832, itself intensified by the fact that there was no municipal sewer system. By then, the Manhattan Company had lost interest in any but the banking aspect of its operations. The reservoir was nevertheless not abandoned until

Remains of the Manhattan Company well, built circa 1800, near Centre and Reade Streets, during excavation for Court House Square building, 1926

Engraving showing the Old Croton Dam, which is now under the waters of the New Croton Reservoir, 1872

1835, and it was not until 1914 that it was torn down.

The first surveys for an upstate reservoir were carried out in 1833. Rye Pond was again considered as a source, along with the Passaic River in New Jersey. But the site settled on was Croton Lake, on high ground in Westchester County, then rural and "a remarkably healthful region," according to a nineteenth-century account.[1] DeWitt Clinton Jr., the civil engineer who headed the surveying commission, proclaimed that "the supply may . . . be considered as inexhaustible, as it is not at all probable that the city will ever require more than it can provide."[2] The reservoir, like all its successors, was to be replenished by rainfall and snowmelt. Construction extended from 1837 to 1842 on a vast project, which

High Bridge during construction of City Water Tunnel No. 1, 1916

included the Croton Dam, creating a 400-acre reservoir; a 41.5-mile gravity-driven brick-lined aqueduct; a 38-acre receiving reservoir holding up to 180 million gallons, known as the Yorkville Reservoir and now the site of the Great Lawn in Central Park; and a distributing reservoir at Fifth Avenue and 42nd Street, on the present site of the New York Public Library. Extending the duration of the project was a protracted debate regarding whether the water should pass over or under the Harlem River. The former won out, resulting in the aqueduct known as High Bridge, the city's oldest extant bridge and a tourist attraction for many years. (The bridge's appeal faded when the river became polluted in the 1960s after construction of the Major Deegan Expressway and Harlem River Drive; it was closed for half a century and not rehabilitated until 2015.) Every element of the Croton project foreshadowed the many New York water-utility projects to come: recalcitrant landowners, aggrieved landowners feeling cheated by their remuneration (some land was bought for $160 an acre and some for $565 an acre), labor disputes, ethnically based disputes among workers, outbreaks of disease, fatal blasting accidents, and innumerable delays.

In 1842, when the Croton system opened, the city's population was 300,000, but by 1860 it had almost tripled. By then, daily water consumption had risen to nearly 70 million gallons, from 12 million in 1842; by 1890 it had jumped to 145 million. The city built a second, larger receiving reservoir, known as Lake Manahatta, just north of the first in 1862—it no longer serves but still exists, known now as the Central Park or Jacqueline Kennedy Onassis Reservoir. Four years later, construction began on the Boyd's Corners Reservoir in Westchester County, and the city started quietly buying up lakes and ponds in Westchester, Putnam, and lower Dutchess Counties—fourteen small bodies of water by 1882. And thus ran the pattern: As the city expanded, more reservoirs were connected to the Croton pipeline. Around the time the population hit the one million mark, the Middle Branch Reservoir opened in 1878; the Kensico Reservoir followed in 1881. In the 1890s, when the addition of Brooklyn, Queens, Staten Island, and the northern and eastern Bronx dramatically enlarged the city and brought the total number of inhabitants to three and a half million, the East Branch (1892), Bog Brook (1893), West Branch (1895), Titicus (1896), and Amawalk (1897) Reservoirs, along with some smaller entities, were added to the chain. By then, the old aqueduct had proved insufficient. The New Croton Aqueduct—which passed under the Harlem River—began construction in 1890 and was finally completed in 1910 (the old one continued to serve alongside until 1955). The New Croton Reservoir, which engulfed its predecessor, opened in 1905.

Water was of course needed not only for homes, but for a wide variety of other establishments. In 1860, Brooklyn listed its commercial and industrial consumers of water: nearly 1,300 shops and stores, more

New Croton Aqueduct pipeline in Central Park, New York City, 1890

Wooden water main, probably installed by the Manhattan Company in the early nineteenth century, unearthed in 1920

than 200 factories, 231 private stables, 136 steam engines, 98 saloons, 70 bakeries, livery stables, hotels, barbershops, machine shops, offices, breweries, markets, churches, fountains, warehouses, slaughterhouses, boardinghouses, schools, jails, hospitals, distilleries, poolrooms, foundries, and refineries. The same day that list was published in *The New York Times*, the paper also ran an editorial, entitled "Cheap Water," which reflected on the fact that the substance was abundant but not limitless, although consumers behaved as though it were and wasted it accordingly. The paper's recommendation was prescient:

> The only reasonable method of preventing waste, is to charge each house with the water which goes into that house, and the only possible method of ascertaining this quantity is *to measure it*, or rather, let it measure itself, like gas, by passing through a meter. . . . [The] price of the difference of water estimated and that really consumed, would in many cases pay the cost of a meter in three months.[3]

At the time, there was little support for metering. Eight years later, one James A. Whitney read a paper at a meeting of the New York Society of Practical Engineering decrying the city's "reckless consumption," pointing out that while London used water at a rate of 42 gallons per day per person, New York consumed 124. He urged the installation of meters. Around that time meters quietly—without press coverage, that is—began appearing in commercial and industrial establishments, although not in dwellings, whether private houses or multifamily buildings. In 1870, someone signing themself Pro Bono Publico wrote a letter to the *Times* warning about the imperfection and expense of water meters, their "grit and sediment," their "frost and friction."[4] When water is no longer free, the tenements will be even dirtier, the writer predicted, and furthermore the supply is increasing faster than the population.

In 1870, too, a new tax levy gave the Department of Public Works the power to install meters in all occupied buildings—previously only hotels, factories, breweries, and distilleries had been subject. By then William M. Tweed, generally referred to as Boss Tweed, had become the city's unelected ruler. He was actually elected to the state senate in 1867 and headed Tammany Hall, the Democratic Party's political engine, but his power came from patronage and vassalage, from serving on numerous boards and commissions, and from maintaining a large network of yes-men in decision-making positions across the political and financial landscape. In 1870, he proposed to install water meters in all buildings at a rate ranging from $125 to $250 per meter—$2,500 to $5,000 in today's money.

Accordingly, he had 60 meters tested at the Croton Yards for three months while issuing no public report. His confederate José de Navarro served as the shill, buying twenty patents and deciding on the Moore meter—which had never been tested, at least impartially. Navarro owned a factory on 22nd Street in Manhattan that began turning out the meters, eventually making 20,000 of them at an estimated value of $18 apiece while charging $50—a profit of $810,000. And that would merely cover stores and other commercial establishments. The profit from apartment houses and the like was estimated at $2.1 million by the water register, J. H. Crane, who blew the whistle.

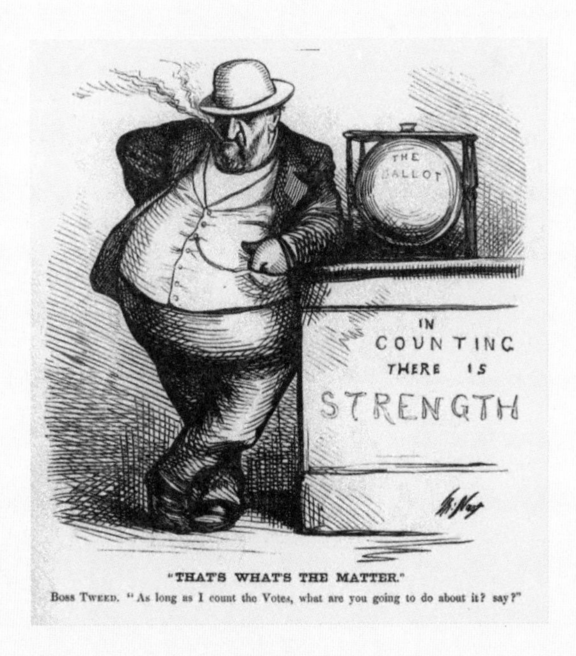

Caricature of Boss Tweed that appeared in *Harper's Weekly* magazine, October 7, 1871

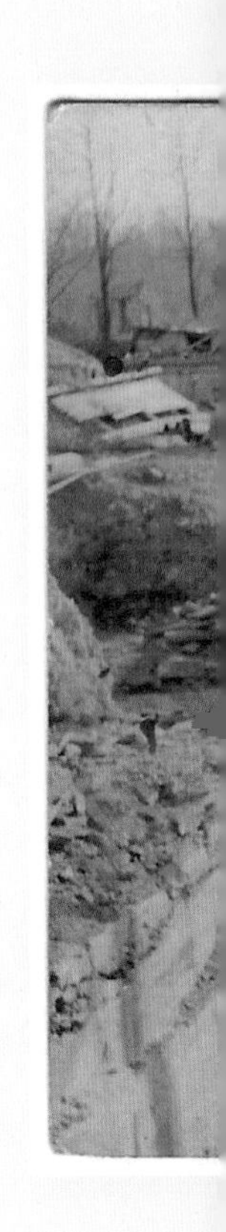

The Moore meter infringed on at least three patents, hence the need to speedily manufacture 20,000 before the lawsuits could begin. Furthermore, the payout was a circular process in which various parties remunerated themselves under other corporate names.

Few responsible people denied the need for meters: The waste of water was inarguable in a time when establishments and private citizens alike ran their taps through the night in the cold months to prevent the pipes from freezing. But Navarro's scheme retarded progress for decades. In 1872, he pressed his claim for $283,500 owed him by the city, and a year later the state supreme court issued a writ of mandamus on his behalf, increasing the amount, with interest, to $416,500. (Meanwhile, the inventor of another meter alleged that in 1870, after his mechanism was tested over a weekend in a locked room, the gauge showed that 2,000 gallons had passed through it—more water than was on hand for the test.) In 1880, the *Times* wrote: "Stacked up in the pipe-yard of the Department of Public Works there are 10,000 water meters, more or less, contracted for by William M. Tweed when he was a power in City affairs, which have never been of any use to anybody under the sun, and for which the City is asked, in the report of a Referee, to pay the immense sum of $1,115,819.19."[5] Said referee, Judge John K. Porter, had served as counsel to Navarro in a matter relating to the elevated railway system. (Tweed, meanwhile, was convicted by a committee of the board of aldermen in 1877 of stealing tens or perhaps hundreds of millions of dollars from city taxpayers. He was jailed, escaped to Spain, was caught and extradited, and died of pneumonia in the Ludlow Street Jail in 1878.) In 1884, the court of appeals ruled that the city owed Navarro $1.5 million with accrued interest, and the following January, the city's controller duly coughed up $1,409,505.65 to Navarro's assignee. In 1891, the meters were finally sold for scrap.

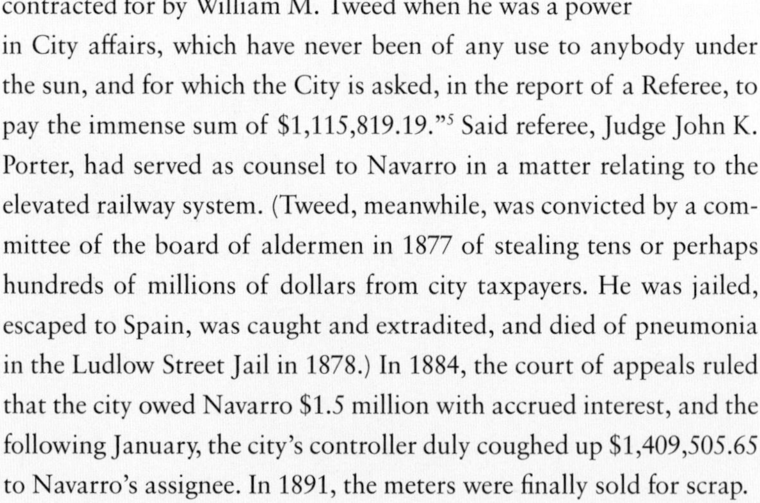

In 1876, water meters were approved by the mayor, the controller, and the chief engineer of the Croton Aqueduct under the provisions

Building the Catskill Aqueduct, postcard, circa 1915

of the Charter of 1878, at prices that ranged from $10.10 for a 5/8-inch meter to $297.50 for a 4-inch model. By 1881, 4,505 meters were in place, 3,261 of them installed by building owners and 1,304 by the Department of Public Works. Water was assessed at a rate of 1.5 to 3 cents per hundred gallons. Matters stalled after that. A letter to the *Times* in 1899 signed Property Owner asked "if it is not about time for the City of New York to place water meters in every house."[6] In 1902, another letter noted that "there is continually a cry to purchase more and more water lands, because in a few years the daily consumption will equal the supply. The present water supply of this city can be made to last at the very least twenty-five years longer without purchasing one acre more of land, by extending the use of meters."[7] A 1903 editorial in the paper advocating

Board of Water Supply police station at Brodhead's Bridge, circa 1915

the use of meters cited the example of Flushing—a part of the city but outside its water system—which could not expand its supply but made up for it with universal metering. After the meters were installed, water pressure increased from 35 to 42 pounds, and total consumption per tap fell from 754 gallons each day to 578 at a time when the rest of the city was consuming 120 gallons a day per capita.

In 1906, the chief engineer of the Department of Water Supply urged meters in every building but met with varied resistance. The borough president of Brooklyn (where consumption was 50 percent greater than in Manhattan) opined that the time would come when New York would have to draw its water from the Great Lakes; a spokesman for the City Vigilance League alleged that half the water in the New Croton Reservoir was going to waste over the dam; the former water register charged that the Department of Street Cleaning was squandering extraordinary amounts of the stuff. Then there were complaints that the official city plumbers were inflating installation prices and forcing kickbacks. In 1907, a justice of the state supreme court ruled against the city in a case brought by a Third Avenue saloon owner over the mandatory installation of meters, his attorney arguing that the order deprived citizens of the right of exclusive possession. Directly under that story in the paper was another headlined "Bids for Ashokan Dam."

A story in 1910 was headlined "Water Saving May Stop Catskill Plan," and attributed to the first deputy to the water commissioner the allegation that 200,000 gallons a day were going unaccounted for. The deputy urged, in addition to universal metering, the installation of valves on the connection to the mains on every block and valves on the feeder pipes to houses, with the aim of stopping leaks. There was also the problem of water theft, with one Bronx firm having allegedly appropriated $525,000 worth in one year, method unspecified. London consumed 36 gallons per day per capita, while New York City swallowed 135. "Every man, woman, and child in New York would have to take four baths a day to make up the difference between the London consumption and our own."[8] (This at a time when London was still New York's model and ideal.) Furthermore, the outstanding debt of $82 million on the Croton reservoir system was a strong encouragement to cut down on water waste before seeking solutions upstate. But Mayor William Jay Gaynor, a reformer who survived a 1910 assassination attempt connected to his efforts to crack down on graft, cited the slogan "Water must be as free as air" and opposed universal metering on the grounds that "it would lead to an economy of water which would be injurious to the health of our community."[9]

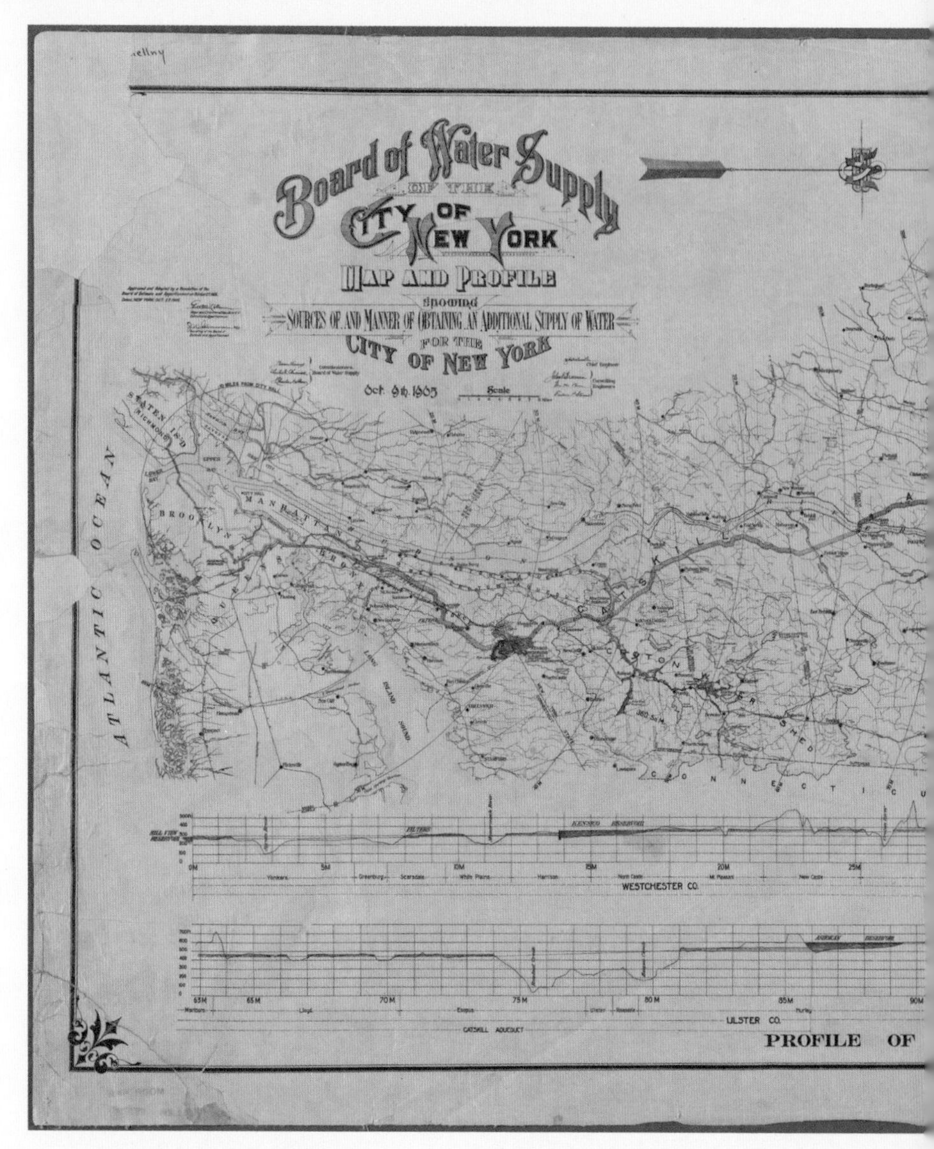

Board of Water Supply map and profile showing sources of, and manner of obtaining, an additional supply of water for New York City, 1905

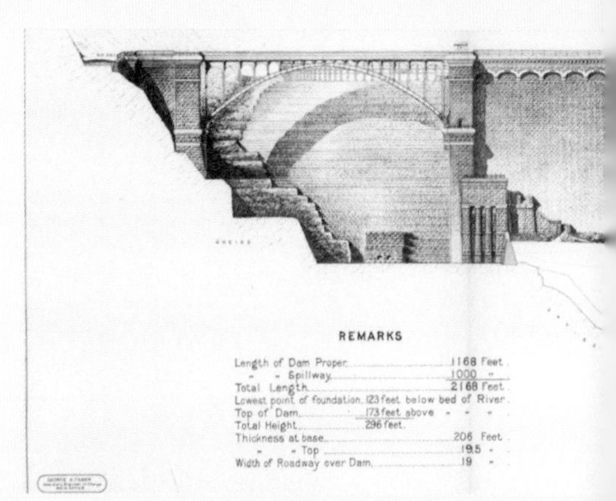

Plan of downstream elevation for the New Croton Dam and Reservoir, 1907

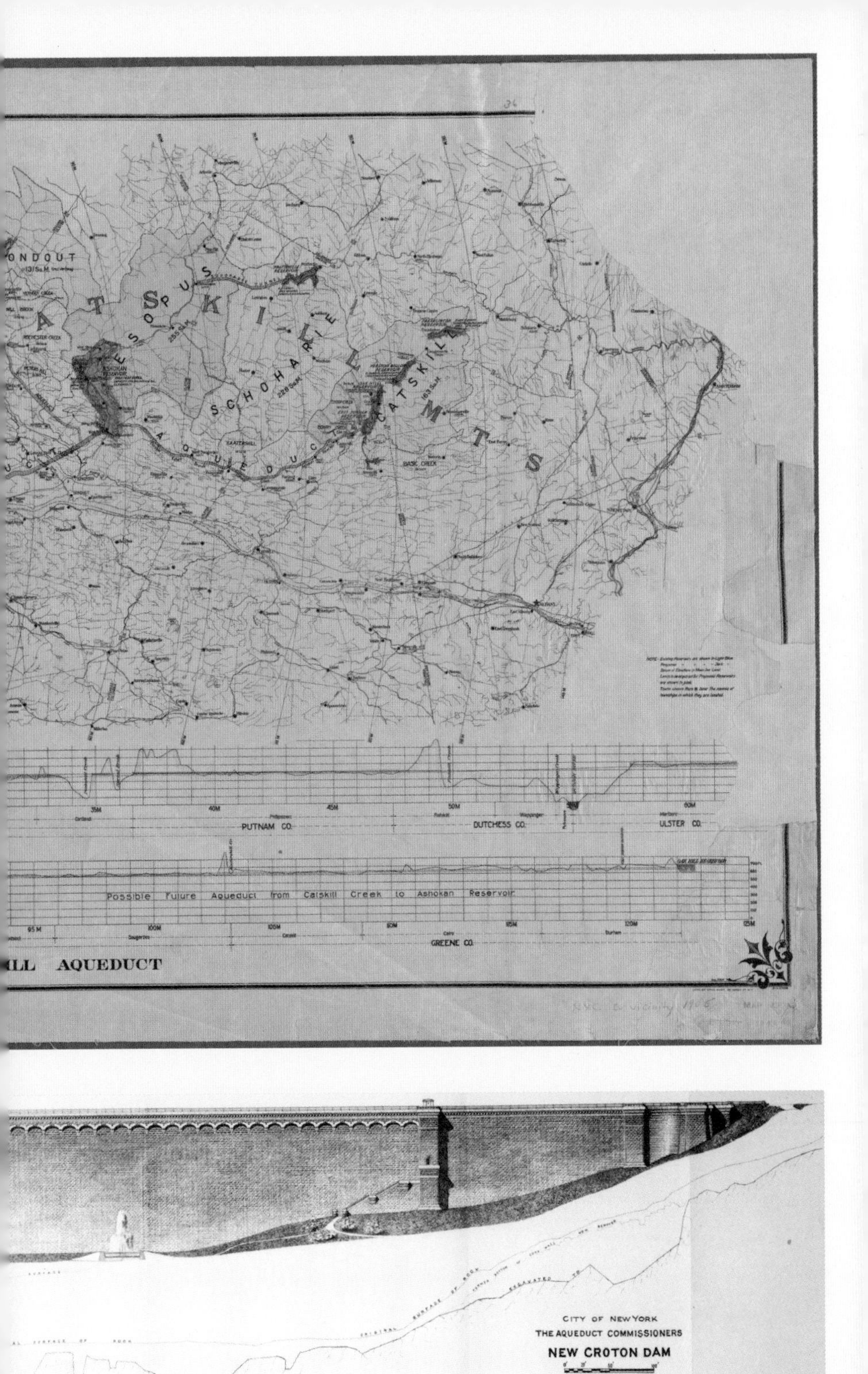
RONDOUT
ESOPUS
SCHOHARIE
CATSKILL
ASHOKAN RESERVOIR
AQUEDUCT
CATSKILL MTS
PUTNAM CO.
DUTCHESS CO.
ULSTER CO.
Possible Future Aqueduct from Catskill Creek to Ashokan Reservoir
GREENE CO.
ILL AQUEDUCT
LIMESTONE
CITY OF NEW YORK
THE AQUEDUCT COMMISSIONERS
NEW CROTON DAM
1906
CHIEF ENGINEER.

The Ashokan Reservoir

Panoramic group
portrait at Olive
Bridge Dam, 1911

At the turn of the century, the city's population had risen to 4 million, and even as the new aqueduct was being completed, it was clear that DeWitt Clinton Jr.'s prediction—"the supply may . . . be considered as inexhaustible"—had been wrong: The Croton system would no longer be sufficient. The idea of tapping the Catskill watershed had first been proposed in 1886, in a paper titled "The Water Supply for New York City" by a chemical engineer named R. D. A. Parrott, published in a supplement to *Scientific American* and expanded by him in another identically titled paper three years later. "The origin of any scheme for furnishing water to a city is found in a desire to supply a demand not yet developed, but foreshadowed,"[1] wrote Parrott, who was known for being the first to derive mineral wool from coal slag but does not appear to have written anything else. Parrott, in any case, accurately predicted what would come to pass twenty years later: a dam on the

Esopus Creek, in the town of Olive in Ulster County on the west side of the Hudson, 100 miles north of New York City. He reasoned that this would be the best location to draw from all across the Catskill Mountains and provided details on the superiority of that range: its ideal ground of sandstone and shale, proving its early geological origin, with quartz crystals indicating that its water table lies beneath any coal formation, along with its exceptionally low quantities of iron, copper, magnesium, soda, and lime.

"Men who have been on the ground and seen the supply say that it will reach anywhere from 300 to 450 million gallons a day,"[2] Parrott wrote, suggesting that he himself had not gone there. Throughout both papers he situated himself at a lofty remove from quotidian human concerns, focusing instead on mercantile ones. He cited the most pressing need for water in the

Cofferdam being built across the Esopus Creek at the Olive Bridge Dam site, 1907

city—$150 million worth of dry goods sitting uninsured in warehouses because the water supply was inadequate for putting out fires—and asked, invoking his own version of the trolley problem, "Under what pretext of dire necessity can it be maintained that one *public* shall sacrifice nine units in order that another *public* may gain one unit?" He warned that "under the beneficent principle of eminent domain [lies] the scepter of oligarchy." But in the next sentence he evaded the matter: "The acquisition of territory under this right presupposes that the use of it shall not injuriously affect adjacent land."[3] He went on to show that the Catskill territory in question suffered from "backwardness" with regard to population increase. Olive Township had 2,924 inhabitants in 1855 and by 1880 had grown larger by exactly three persons; the total affected watershed population was 23,430 in 1855, and in 1880 it was 23,436. The combined population increase in Ulster, Greene, and Delaware Counties between 1870 and 1880 was one-and-a-half percent, while in the watershed townships it was minus seven-eighths of a percent. Furthermore, Parrott assured his readers, "It is assumed as a generally admitted fact that New York City exerts a positive influence of growth on the surrounding territory."[4]

There were other contenders besides the Catskill counties, yet those faded as soon as the city began its inquiries. In 1904, Dutchess County (on the east or "Croton system" side of the Hudson) passed an act banning city access to its Ten Mile River watershed or any other source. The prospect of a similar ban scotched an idea to tap the Housatonic River in Massachusetts, while the perennial notion of drawing on the Hudson itself proved perennially unpopular given the river's pollution—despite the fact that the city of Albany drank filtered Hudson water.

In 1905, New York City's Board of Water Supply assembled a crack team of engineers and sent them off to survey the Catskill territory: J. Waldo Smith, the chief, who had become chief engineer on the New Croton Reservoir in 1903; John R. Freeman, who had consulted on the Panama Canal and the Grand Canal in China; William H. Burr, professor of civil engineering at Columbia University; and Rudolph Hering, a German-born sanitary engineer. They recommended what

Parrott had envisioned: a dam on the Esopus watershed in Ulster County. After conducting drilling tests at various points—Cold Brook, Lake Hill, Wittenberg, Shandaken, Big Indian—they proposed that the main dam be constructed at Bishop Falls in Olive Bridge, which had solid bedrock and would permit the work to be done two years sooner than at rival sites. Ulster County did not possess sufficient political sway or capital to resist.

Bishop Falls was a tourist attraction, a gently curved array of shallow falls in narrowing tiers like a fan, shown in numerous prints and postcards poised between a mill on one bank and a barn on the other with Catskill peaks in the distance. The area had long been a summer resort for city people who could afford second homes but had more recently been democratized by the establishment of boardinghouses, where visitors would pay for a bed, perhaps a bath, and three meals taken communally, and generally spend their time loafing. John Boice, who owned the falls, valued his property at $500,000. Under the relentless terms of the city's condemnation commissions—the three-person boards that reviewed evidence, heard witnesses, and set final valuations on land the city sought to seize—Boice was eventually given $112,303.18. Bishop Falls now lies at the deepest point of the reservoir. Isaac Whitaker owned the "romantic gorge" below the

Bishop Falls, postcard, circa 1905

falls, which was said to be "one of the most sightly places in the mountains"[5] and was constantly subject to offers by prospective hoteliers; it was assessed by the city at less than $5,000.

In those days the *Kingston Daily Freeman*, the most important newspaper in Ulster County, had correspondents in villages who would report primarily on social events in dedicated columns. In March 1905, the paper's man in Shokan was moved to editorialize:

> Many Shokanites are wondering at the new and evident opposition on the part of some Kingstonians to the proposed New York City Ashokan reservoir. Why citizens of Olive are and should be, some angry, some terror-stricken, some confused, all can understand. But that Kingston and Ulster County should antagonize the project is more of a puzzle to us. We never dreamed that it was true, as is now heralded abroad, that our little valley contained such a large percentage of Ulster's taxable property, the most of its fertile land or that our population was so enormous and our trade the chief asset of Kingston's merchants, and our destruction the greatest calamity that could befall the Empire State. Mayhap if the water does not come we will hold our heads higher hereafter in consequence of a true appreciation of our importance, and perchance, if the water does come our residents will be more equitably reimbursed for their loss of homes and seizure of property, since publicity has truly appraised us.[6]

And then he went on to note that Landon Churchwell was ill with pneumonia, that Mrs. Abigail Markle was visiting her daughter in Brooklyn, and that Miss Inez Dumond had returned safely from a visit to Port Ewen.

Despite such attentions from the newspaper, the Ashokan Valley was not in fact highly regarded in Kingston, the county seat, where more prosperous citizens regarded themselves as urbane, cultured, and educated to a degree that made them equal to their metropolitan cousins and quite unlike the hayseeds of the valley, who were barely capable of subsistence farming and added little to the county's economic profile.* Despite its farms, which seldom produced much to export; despite its bluestone quarries, which for decades had supplied sidewalk paving to the city of New York; and despite its forests, rich in spruce, pine, maple, ash, oak, and hickory—which Parrott dismissed as merely providing material for "chairs, piano bars, roller blocks, clubs, bowls, trays, etc."[7]—the valley's principal economic engine

* "Ashokan Valley" here refers to the land seized for the reservoir, as distinct from the much longer course of the Esopus Creek, which begins on the slopes of Slide Mountain and pours into the Hudson River at Saugerties.

West Hurley, circa 1899

Olive City, circa 1899

Shokan Village and the Ashokan Valley, with Ulster and Delaware train, circa 1899

was in fact tourism. (Even so, Parrott wrote: "To exclude the boarder and the excursionist from 530 square miles out of 4,121 would not be a public privation."[8]) It is fair to say that the valley was taken for granted even by the local powers, who in that time of accelerating industrial progress viewed it as a mere beauty spot, permanently unimprovable.

The territory acquired by New York City between 1905 and 1910 was a strip measuring roughly one by twelve miles, a total extent of 8,300 acres. Included were the villages of West Hurley, Glenford, Ashton, Olive Branch, Shokan, West Shokan, Boiceville, Brodhead's Bridge, Olive City, Olive Bridge, and Brown's Station. Also seized were the tracks and right of way of the Ulster and Delaware Railroad and some parcels of forest preserve owned by the state, which later led to rapid legal adjustments when statutes turned up prohibiting these woods from being flooded. The inhabited area contained 504 private houses, 35 stores, ten churches, ten schools, nine blacksmiths' shops, seven sawmills, and a gristmill. The $162 million earmarked for the Ashokan project included the cost of acquiring these properties, along with the construction of the dam, the dike, the waste and dividing weirs, the embankments and roadways, the temporary railroad that served construction, and of course the aqueduct, 92 miles of cut-and-cover, grade tunnel, pressure tunnel, and steel-pipe siphon, traveling under the Hudson near the mouth of the Wallkill River to connect with the Croton watershed at the Kensico Reservoir, back on the east side of the Hudson, outside what is now the New York City suburb of White Plains.

This was all made possible by the McClellan Act of 1905, named after the then mayor of New York, George B. McClellan Jr.—son of the Civil War general, he was a conservative Democrat later famous for canceling movie house licenses on Christmas Eve 1908 as both a moral and a fire hazard. Among other things, the McClellan Act "allowed the city to take possession of land and/or dwellings just 10 days after an appraisal commission was appointed and upon payment to the owner of one-half the assessed valuation of the property."[9] Charles T. Coutant, who represented Ulster County in the New York State Assembly, introduced a countermeasure, the Coutant Bill, which argued that the city could very well take its water from the Hudson, since Albany did so, and could put its land-purchase money into a filtration system instead. At a hearing in Albany, another Ulster County representative, Samuel D. Coykendall, warned, "If our mountainsides and valleys are ever taken by New York City, it will not be with the consent of our people, but because it is forced upon us by the representatives of the other counties in the state. And please remember, gentlemen, you who represent interior counties, if Ulster County is sacrificed your turn may come next."[10]

The *Kingston Daily Freeman* reported in May 1905 that experts appointed by the city to the condemnation commissions were instructed to accept only a third of the valuation placed on land by its owners; a local expert denied he had been told to fix a value of $20 per acre across the board on all condemned lands. The total value for the valley's residential area (minus, that is, land owned by the railroad) was $34 million; shortly after this figure was made public, an item ran in the *New York World*: "A silver ewer has been sold at auction in London for $21,000. In proportion to the amount of water it will hold it was relatively as dear as the Catskill reservoir."[11] At a hearing before the state legislature, when the county demanded protection from the land grab, the city's "attitude was best exhibited by Corporation Counsel [John J.] Delaney when he assured the Ulster County property owners that they need never fear about collecting their money for they could come to New York and 'attach City Hall.' [. . .] 'Cheer up, boys, there ain't no hell,' was the quotation he used in a peroration filled with pathos—and wind."[12] While Delaney was sardonically placating upstate landowners with a fantasy of seizing city hall in a lawsuit, an appreciative crowd at the Kingston Opera House took in Billy Bitzer's one-reeler *In the Valley of Esopus*. Bitzer—the cinematographer who would go on to shoot many of D. W. Griffith's movies, and who is credited with introducing, among other things, the close-up, the long shot, the fade-out, the iris, and artificial lighting—filmed his picture in a single take from the rear platform of a train traveling backward through the valley on the Ulster and Delaware tracks. It featured a bit of plot: A man fishing off a trestle refuses to move and the train has to stop, whereupon two other men appear from nowhere and throw him off the trestle. The film now appears to be lost.

The actual process began in the spring of 1907, when the first of the nine condemnation commissions got to work. These commissions were formed sequentially, although their actions overlapped. They seized land just before it was needed for the process of demolition and construction, so that, beginning at Bishop Falls on the southern tip, they moved first eastward and then clockwise through the valley, closely followed by the contractors, in a process that would not end until 1911. In the summer of 1907, *The New York Times* found the doings rustically comical: A farmer told their reporter he wanted the

Ceremony inaugurating construction on the Catskill Aqueduct near Indian Brook, 1907

city to pay for rabbits and woodcocks driven away by the commotion; someone else's cow ate a stick of dynamite and died, for which he was given $100 by the contractors. But, of course, each condemnation revealed a small, specific narrative. The local jumble of small villages and smallholdings was home to other recurring types—small shopkeepers, boardinghouse keepers, millers and blacksmiths and quarrymen, ministers and schoolteachers. Their connection to the world beyond their own fence lines was, at best, the daily newspaper; few had

Road construction atop the middle dike at the Ashokan Reservoir, 1915

telephones. The trip to Kingston, no more than ten miles away, was undertaken by most maybe a couple of times a year. They spent very little money, consuming mostly things they cultivated themselves or bartered with a neighbor. But they weren't isolated; dense family ties spread across the region, and neighbors had been the same for decades. They were bonded and pledged and sealed to the land. "There probably are not two properties exactly alike in the entire Ashokan region," wrote the *Kingston Daily Freeman*, "and elements and circumstances which enter into one case will be supplanted by entirely different elements and circumstances in the succeeding case."[13] Recurring names in the newspaper stories show descent from the Dutch, who had arrived in the area around 1614: Boice, Elmendorf, Krom, Van Kleeck, Bogart, Stoutenburgh, Van Wagenen, Palen, Winne, Roosa. "Among some of the people of Ulster County," an engineer on the Board of Water Supply told *The New York Times*, "many deeds are unrecorded and have to be obtained directly from the owners to be copied. In the mountains deeds are handed from one person to another when land is transferred, and so pass through many hands without ever being recorded."[14]

Charles Pierson's farm was exemplary. A bachelor at age 52, he owned 23 acres inherited from his father and grandfather that had been improved with a house, a hay barn, a cow barn, a granary, a henhouse, a woodhouse, twenty acres of meadowland, an apple orchard, a plum orchard, and a raspberry vineyard. He had harvested 33 bushels of oats and two tons of hay per acre for almost twenty years (the average price per ton at Olive Branch Station had been $15 for the preceding five or six years); he grew potatoes, onions, corn, and numerous smaller crops. His location was ideal: three-quarters of a mile from the train station and eight miles from Kingston. He lived alone, but probably hired help at harvest time and had siblings and their families close by; he had run his farm for decades according to his own terms and on his own schedule. The records do not say what became of Charles Pierson—except, presumably, to note his compensation in a long list of settlements. He may well have purchased another farm west or south of the reservoir, but this could never have been the same as what he had lost. There are no accounts of illnesses or deaths, emotional crises or suicides following the land seizures, and the term "post-traumatic stress disorder" was decades from being coined. But such sufferings must surely have been common. Each determination by the condemnation commission in effect dismantled a life—and at cut rate.

The most famous case, locally, was that of Rachel Everett of Olive, who claimed $10,000 for her house, land, and ginseng garden. The court needed to learn what ginseng was, which led counsel to pronounce the word "aphrodisiac," leading in turn to three days of bantering between lawyers and many more of newspaper witticisms. Everett's husband, Egbert, testified that he had been born a few miles from the couple's home in Olive and had been engaged in the trucking business in Brooklyn, but that a degenerative spine disease had caused him to become almost completely blind. He had discovered the lucrative properties of ginseng, and with his wife had bought the land and undertaken its cultivation. He testified that North American ginseng went for between $6.50 and $8 per pound, whereas Manchurian ginseng fetched $40 for the same amount. The Everetts sold their produce to the Mee-Wun-Wing-Lee Company, which exported it to China; Egbert alleged that all the ginseng produced in the United States since 1852 had been similarly exported. Their counsel tried to support their claim by arguing, "There is an unlimited demand, as there are over 400 million Chinese."[15] When the city granted only $2,900 to Mrs. Everett, disregarding the value of the ginseng business, the court intervened to award her $8,000. But this was an exceptional case.

The first awards in damage cases came in November 1907. The total claimed by Ashokan Valley property owners was $122,610.30; the city's total valuation was $44,731.49. The court awarded $50,600, with an additional $4,086.18 in counsel and witness fees.

The fees racked up by lawyers, appraisers, and expert witnesses, however, were bounteous—so much so that the *Freeman* and other papers began referring to the proceedings as "shaking the plum tree" and the Ashokan region as "plumland." Prices were negotiated per counts of individual trees and vines in orchards, but stalled on usage values for any kind of land, a matter that came down especially hard on owners of the many small bluestone quarries, as well as the proprietors of the boardinghouses, who were compensated only for their buildings' value and not for lost business. In July 1907, Appraisal Commission No. 2 ruled that property owners could not recover damages to their land—a stratagem apparently meant to compensate for the commissions' treating all parcels as equivalent whether they held farmland, brush, a quarry, a dump, or a natural prospect—or be remunerated for the loss of business conducted upon it, and the ruling was upheld by the special counsel. When Commission No. 5 attempted to counteract the ruling, accepting evidence of damage and lost business, the special counsel objected, "declaring, according to the report, that for a man whose property was taken to get damages for his business was 'absolutely monstrous.'"[16]

The most persistent litigant was Mrs. Tina Lasher, who ran a boardinghouse and a small store near the depot in Brown's Station, with which she supported herself and her children. She claimed $18,712 for her 24 acres, including $3,000 for loss of business. Her witnesses testified to a value of $13,212.50; witnesses for the city countered with a value of $7,500; in July 1907, the city allowed her $6,500. Mrs. Lasher was denied permission to testify before Commission No. 2, but Justice Cantine of the state supreme court granted her an injunction and order to show cause, requiring the city to demonstrate why a permanent injunction should not be issued restraining them from interfering with her business. In April 1908, state supreme court justice Howard denied the permanent injunction but ruled that business owners in the Ashokan district were indeed entitled to recover damages. In December, Mrs. Lasher's attorneys served papers on the mayor of New York, the Board of Water Supply, and the corporation counsel to compel the city to appoint a commission to ascertain damages to her business. In February 1909, Justice Howard granted her a writ of mandamus to further compel the city in the matter. "Every owner of a business in the Ashokan section received the news as a sure indication that claims for business and indirect damages would have to be recognized and settled by the city."[17] The city took the case to the Appellate Division of the

Supreme Court, which handed down its decision in September 1909, upholding Howard's decision. In May 1910, the court of appeals also upheld the verdict. Before this occurred, however, Mrs. Lasher died; the date and cause are unrecorded and there was no obituary. Nevertheless, from that point—a bit late in the game—the city had to recognize claims and appoint commissions to determine business loss.

Meanwhile, construction was proceeding. The Olive Bridge Dam was to be 220 feet high, 5,000 feet long, and 26 feet wide at its top. Its masonry was of the kind known as "cyclopean"—a type of ancient Mycenaean stonework using massive irregular blocks—which involved dropping huge chunks of uncarved local stone into beds of concrete. The dam would be built in 83-foot sections to prevent cracking from contraction and expansion caused by weather. The ends would be earth embankments 700 feet thick at the base and 34 feet at the top. To prevent seepage, a system of drainage wells would be sunk twelve feet apart in the dam's upstream face and connected to inspection galleries. Along the axis,

Test shaft at Breakneck Ridge during construction of the Catskill Aqueduct, 1907

a core wall of concrete would extend to the bedrock. The dividing weir would feature a causeway accessible to vehicles, from which one could enjoy the "lacustrian" scenery. The whole undertaking would require the excavation of around two million cubic yards of dirt and 400,000 cubic yards of rock. It would call for 1,100,000 barrels of cement for 900,000 cubic yards of masonry; the embankments alone would require 7 million cubic yards of material—enough to build a pyramid 200 feet square at its base and 400 feet high. The grandeur of the project—and those that would follow it—was invariably expressed in figures. That was the poetic language of enterprise in the twentieth century. Nothing else conveyed so well the immensity of every new undertaking and its dwarfing of whatever had preceded it, and it could be appreciated by the average piker with a fourth-grade education. Nowadays numbers have become so large they usually dissolve into abstraction.

By September 1908, a substantial camp was being constructed at Brown's Station for a population of some 6,000 diggers, carters, foremen, and engineers, most of them unskilled, "including negroes, Poles, Italians, and Slavs."[18] The settlement was intended to last just seven years and was accordingly never named, but streets were laid out with 125 four-room cottages, dormitories, a hospital, a bank, a company store, a sewage plant, and even its own reservoir. Still to come were churches and schools—the latter overseen by the combined forces of the Board of Water Supply and the Society for the Protection of Italian Immigrants—and a police force intended to curb petty theft and quell disputes, especially between ethnic groups; this corps survives today as the Department of Environmental Protection Police Force, which has jurisdiction over the Ashokan Reservoir and surrounding property owned by New York City. In effect, the workers' settlement became a city, with all a city's problems: assault, theft, and murder, not to mention drunkenness in significant quantities. There was no saloon on-site and no liquor sold at the company store, but dives, often with a sideline in prostitution, sprang up all over. At one point, there were said to be fourteen on the road to Marbletown alone. As for the police, the *Freeman* wrote, "The men composing

Both photographs of camp schools established near the village of Brown's Station for the children of Ashokan Reservoir workers, 1908

the new police force have been taken from various trades and businesses, and in some cases they and work have been strangers. They are not supposed to know criminal law and they don't."[19]

Eventually there were "musical and dramatic entertainments and illustrated lectures,"[20] including theatrical productions in Italian. But the dam workers' city seems to have been a predictably tough place, rife with tensions between Italian Americans and African Americans and Eastern European immigrants, many of the latter conscripted abroad by agents of the contracting firm building the dam and shipped upstate directly from Ellis Island with zero knowledge of English. There were nearly daily conflicts between the camp's administrators and peddlers who came in to sell meat, bread, and milk of higher quality and at lower prices than what was available at the company store. And injuries, some fatal, were frequent on the work site: falls, clothing caught in machinery, a pickaxe hitting a stick of dynamite. Still somehow the work got done on time, and the Ashokan Reservoir opened in 1915 and quickly became a tourist attraction. But even then it was clear that the new Reservoir would be insufficient; newspapers were discussing the proposed Neversink Reservoir as early as 1911. This process—of anticipation or second-guessing—would continue for another half century.

The communities that were to be eliminated, some of them dating back to the mid-eighteenth century, were valley villages of the sort that, had the reservoir never intervened, might have grown to support two gas stations and a regional high school—or they might have just blown away, their only remaining trace a historical marker on a roadside. At one commission hearing, a witness from New York City professed himself unable to construe what, exactly, was Brown's Station. A citizen named Edwin Burhans volunteered to enlighten him.

Ashokan Reservoir camp hospital, October 30, 1908

Pest house for isolating reservoir workers with communicable diseases, 1909

Ashokan National Bank and post office in Brown's Station, postcard, circa 1909

> Brown's Station includes everybody who receives mail from the Brown's Station postoffice. There are fifty-seven buildings. Twelve of these are used for residential purposes only. There is a school maintained by the local authorities, a postoffice and a railroad depot. A public telephone, telegraph office and a store complete the places of prominence.[21]

He added that from the post office the village extended one mile west, more than a mile east, one-and-a-quarter miles north, and a mile-and-three-quarters south. The population consisted of fifty-seven families; there were nine or ten boardinghouses, two sawmills, a pulp mill, and several quarries. There were five roads, one of which led to the pulp mill and then went on to Kingston, the others with more local issue. Pressed by a commissioner, Burhans admitted that, yes, all five roads could really be considered one big road.

Besides the living residents, the villages also harbored the dead, and these, too, had to be moved. There were 2,413 graves, or 2,637, or

2,720, or 2,800—every source gives a different oddly specific total—in a dozen cemeteries and many more small family graveyards, one of which turned out to abut a much older burying ground established by the Lenni Lenape at an unknown date. The Native American remains, along with unclaimed or unidentifiable bodies of European Americans, were removed to a cemetery at the far end of Watson Hollow, west of the westernmost part of the reservoir. There had been a tussle over damage claims to monuments before a judge ruled that tombstones were legally real estate, and hence that families could be compensated for their loss or relocation. It cost a total of $300,000 to move the bodies, the sums mostly borne by their descendants: $42 for a new lot in the town graveyard of their choice, four dollars to open the grave, four dollars for reinterment, $20 for a cheap metal coffin, ten to transport the remains, five to move the headstone. The city allowed $15 for the move and $3 for the stone.

Once New York City had acquired a property—by 1909 it owned all the land in the Ashokan Valley—it would clear the area immediately if it stood in the way of imminent construction. Most holdings, however, were not urgently required at first, and so the city would rent out the houses, sometimes to their original inhabitants. It would auction off produce that did not need to be tended, such as apples (in 1911 the city made $457 on the apple harvest), and reap the hay, which would be fed to police horses and labor mules. Reminiscing in the 1980s, a woman who had been a small child during construction of the dam remembered that a mule would begin to bray every morning at exactly ten minutes to five, soon accompanied by all its stablemates—a more effective wake-up than the company whistle.

Three hundred animals were employed in the project (331 at its peak in August 1909). The mules, who accounted for about 200 of the total, worked in teams of three, driven by African Americans, most of whom had been recruited in the South, mainly Alabama. These drovers, who in one account numbered 123, lived in a segregated camp, allegedly for their own protection. But then the other camps in the great temporary city near Brown's Station were segregated as well. The workers on site numbered between 1,900 and 2,500 depending on

View of the Ashokan Reservoir from the
J. Waldo Smith Monument, 1926

ACC. HDQ.R. 2059. 9.30.26 .3.

Drovers and mules at the reservoir site, circa 1910

the phase of the project (there were also 178 women, primarily there because their husbands were, and 301 children). More than half were Italians—most unskilled, but some of them master stonecutters—and their camp was called Little Italy. The papers spoke darkly of the so-called "padrone" system of patronage and kickbacks; any time an Italian was arrested, he was alleged to be a member of the Black Hand. Other workers came from Austria, Poland, Russia, Sweden, Finland, and Germany, and naturally there were some locally recruited hands as well.

Highest paid were the stonemasons, at $3 a day. Pipefitters, pumpers, and plumbers got $2 a day; powdermen, who worked with dynamite, earned $10.16 a week; unskilled laborers received between $1.20 and $1.60 a day. Waterboys, as young as nine or ten years old, were given a dollar a day for many mile-long round trips bearing two twelve-quart pails. From their wages workers paid for room and board at a rate of $20 to $22.50 a month. About a hundred laborers went on strike in 1908, demanding an increase of five cents a day, and the Industrial Workers of the World sent out an organizer, but the strike failed and the workforce was never unionized. The jobs were hazardous; workers were killed or maimed by collapsing scaffolds, inopportune explosions, being scalded by steam, being pulled into stone crushers or caught in the gears of steamrollers—and subject to lesser injuries and disease, such as the typhoid outbreak of 1913. But the city was inconsistent in its care. I was unable to find even a speculative statistic for on-the-job fatalities.

The *Kingston Daily Freeman* covered many of the accidents, sometimes in grisly detail. It also recorded on-site murders, of which there seem to have been maybe half a dozen, all of them accomplished with knives, and it published summary notices of raids on dives and brothels and numerous arrests of illegal hooch vendors. By and large, however, its focus was on the legal and financial machinations of the project. There was the ongoing matter of the local telephone companies, of which there were two: the Hudson River Telephone Company, which received a settlement of $25,000, and the Citizens' Standard Telephone Company, which got $15,500, both of them also granted perpetual use of New York City's right of way for their cables. However, the companies had paid easements of a dollar apiece to all the landowners whose property their cables originally traversed, and these were never recompensed by the city. More serious was the long struggle between New York City and the Ulster and Delaware Railroad. Its owner, Samuel D. Coykendall; its general manager, his brother Edward; and its general counsel, Judge A. T. Clearwater, had been among the earliest and most outspoken opponents of the Ashokan Reservoir and had made significant efforts to defeat it through the legislature and the courts at Albany. They were prepared to fight it out to the end. Faced with defeat at the hands of the city they demanded $4,555,174, including $1,600,000 to relocate the tracks on a route just north of the reservoir; in June 1911, they finally settled for $3.1 million in compensation, with a $200,000 rebate to the city if it laid the new tracks for them. Two years later the route had been built, minus

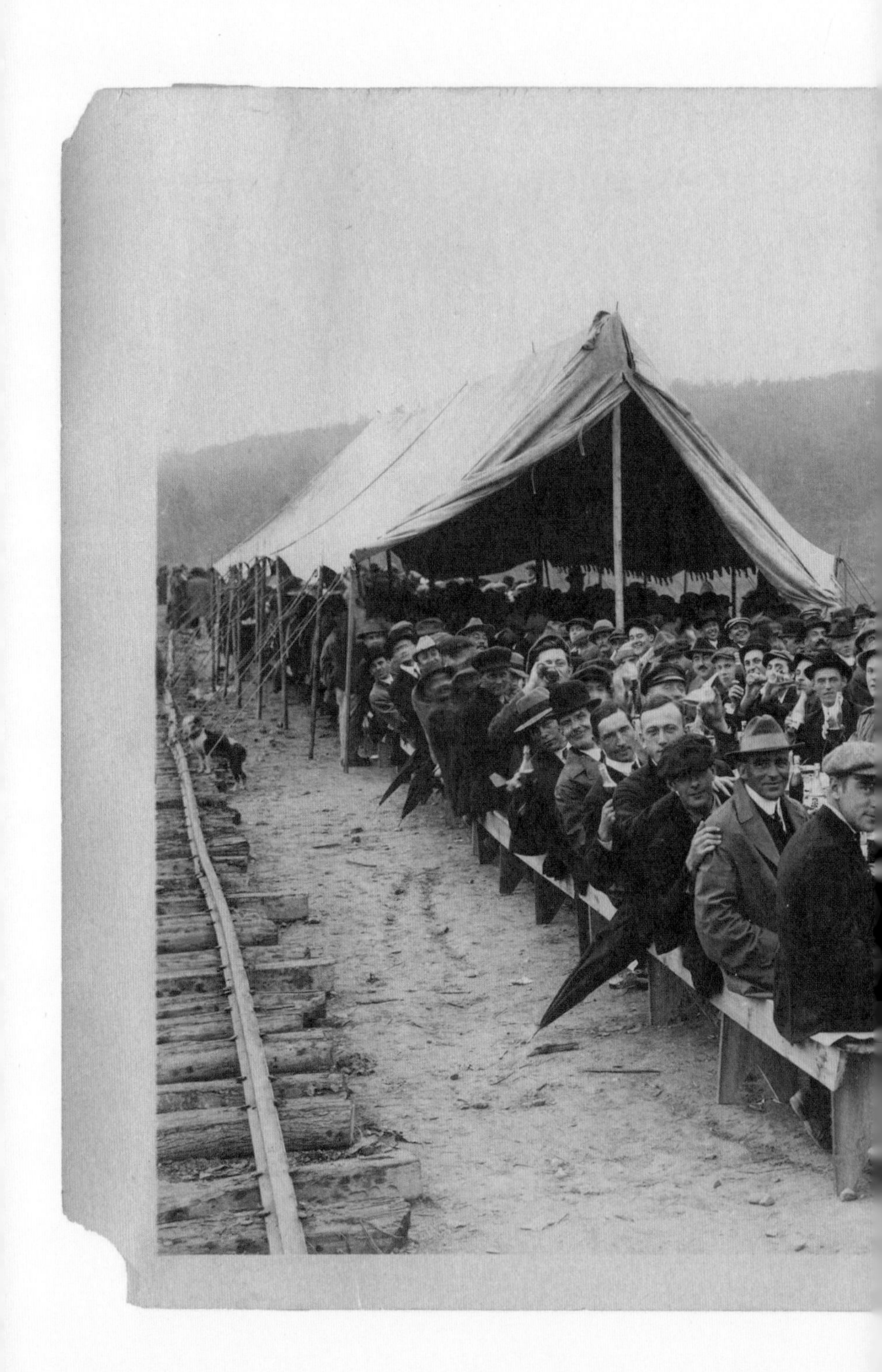

Lunch atop the Olive Bridge Dam, with fried chicken, apple pie, an Ashokan cocktail, and songs, 1913

Ulster and Delaware train at the Brown's Station depot, circa 1899

six stations but including two new ones, and rolling stock shifted to it, soon to run at least twenty trains a day. On opening day, Curtis H. Peters, the reservoir project's chief mechanical engineer, was killed when he drove his automobile over the old track, believing that all trains had been suspended. All work stopped for his funeral.

From 1907 to 1913, a steady trickle of people moved out of the Ashokan Valley. Of 1908 migrants, 163 died during the six years of the project, 712 relocated in the immediate vicinity of the reservoir, 472 moved to Kingston and its surrounds, 71 moved west to Woodstock and Mount Pleasant, 162 moved south to Port Ewen and environs, and 328 moved clear off the map. Sometimes entire communities were relocated. West Hurley, Shokan, West Shokan, and Boiceville shifted themselves north or west, moving buildings as well, even if it meant repurposing them in the new setting: A church became a fraternal hall, a school became a store. A 1913 photograph of relocated Shokan looks much the way it does today along a straight stretch of Route 28: a widely spaced row of substantial houses—presumably carted there by their owners on log rollers, drawn by oxen—across from the the Ulster and Delaware tracks, nowadays idle. The only village to relocate south of the reservoir was Olive Bridge. Although Glenford did not relocate, the congregation of its Methodist church decided to move its edifice north. The only wrinkle was that part of the building lay on the newly designated Ulster and Delaware line, which gave the railroad majority ownership. Coykendall had no objection to the move, but deferred salvage rights to the reservoir's general contractor, J. O. Winston, who approved the shift at no charge. So the church was moved to a lower slope of Ohayo Mountain (where it still stands)—whereupon the city demanded its immediate return, on byzantine legal grounds. The case finally went to trial in December 1914, the press rejoicing in the phrase "the stolen church." The court at length returned a sealed verdict awarding the city $45.

Money was an endless source of discussion as reports were released, sums contended, lawsuits joined; the city was still negotiating individual settlements late in the 1930s. The largest single award to a private citizen was made to John Boice, owner of Bishop Falls, once a celebrated beauty spot and now the site of the main dam, who received $112,303.18, with $27,000 in accrued interest and $12,000 in costs. By 1910, the property awards amounted to nearly $4 million—and the cost of the attendant proceedings almost $3 million: 75 cents on the dollar. That year, an editorial writer for the *Freeman* brought up a current scandal concerning lawyers who had descended on the Oklahoma Territory and were extorting unwarranted fees from Native Americans who were trying to get proper compensation for lands they sold to the government in the passage to statehood. Much the same was occurring down the road, the *Freeman* observed.

> Perhaps they [in Oklahoma] had heard of how civilized white men in Ulster County had done very much the same thing, agreeing to pay a bonus of 10 per cent to a lawyer whose pay was already guaranteed by the city of New York, or believing the tales they heard and selling their property for less than its value to speculators who hoped to collect many times its value.[22]

Those speculators, otherwise unrecorded, must themselves have felt cheated when the payout came. The city did very well. Its completion estimate had been $176,663,000; the final total cost was $184,707,540, the difference accounted for by the need to expand the reservoir police and an enlarged workforce owing to the demands of the newly won eight-hour day. Nevertheless, the city was known for not paying its bills. The prosecution of crimes committed by reservoir employees had strained local budgets to the tune of $13,000 by 1910, but the city would compensate local authorities only if the crime had been perpetrated by an employee on a day of employment—it declined to cover Sundays and absences and would not pay justices of the peace and the like. In 1910, the city began a campaign to make the reservoir tax-exempt, which failed. Piqued, the city refused to pay its tax bill for years, until in 1930 the Ulster County treasurer and the supervisors of the townships of Olive, Hurley, and Marbletown put the reservoir up for sale for back taxes. The city finally coughed up $576,023.40, with $54,000 in interest and fees, in 1932.

By June 1913, the last stubborn holdouts moved out of the valley—though more than 800 had still been living within the taking lines earlier that spring. The place was eerie by then, with few recognizable landmarks. Lifelong residents of Brown's Station and Olive City would get lost in their own villages and need to ask their way. Workers were still blasting stumps and burning trees, and the night sky was punctuated by columns of fire. (Despite the long-running legend that droughts afford a chance to see steeples cresting above the reservoir's waterline, all remaining buildings had been burned as well, and even the ashes removed; a severe drought now reveals only cellars and roadbeds.) In October, an inspection by the New York Board of Water Supply indicated that roughly two years' worth of finishing

Traver Hollow Bridge under construction at the northwestern end of the Ashokan Reservoir, 1913

touches still needed to be done. But water was now flowing into the excavation, the reservoir "so nearly completed that were New York City to be afflicted with a water famine within the next few weeks the Catskill water supply could be turned into this immense connecting tunnel and from the Croton watershed sent to the city through the regular channels."[23]

In 1914, the wings of the dam were paved with three-to-four-ton blocks of bluestone quarried locally, and 780 men and 244 mules and horses laid macadam over the 40 miles of reservoir roads. The local water was so pure that (as the aptly named Judge Clearwater pointed out in 1905, in a speech against the proposed reservoir) New Yorkers had long been buying it in five-gallon carboys from the Crystal

O. DuBois

Top: Ashokan main dam under construction, 1911

Middle: Men on horses atop one of the dikes built to hold water during the construction of the Ashokan Reservoir, 1911

Bottom: View of the Ashokan Reservoir from the J. Waldo Smith Monument, 1916

Headworks and aerating fountain at the Ashokan Reservoir, 1917

Spring Water Company of Pine Hill. So the facility did not need a purification plant, but was instead given an aerating fountain. Water passed through an aeration basin, a small reservoir 500 by 250 feet, its bottom lined with pipes four or five feet apart, from which streams were ejected 40 to 60 feet in the air. This oxidized vegetable organisms and removed taste and odor. (Eventually alum was introduced to counter the effects of turbidity and soda ash to prevent over-acidity, and the water was chlorinated twice at the Kensico and Hillview Reservoirs.) On June 24, 1914, all the steam whistles in the zone were set off at once, marking the official end of the project. Only the cleanup remained. Throughout 1916, workers slowly demolished the camp at Brown's Station and the plants and temporary railroads.

The year 1914 had been dry, but in 1915 the rains came, filling the reservoir to a hundred feet; on November 22, water was released into the aqueduct. In an era of superlatives, when postcards were published comparing, for example, the height of skyscrapers with the length of ocean liners, the press reveled in statistics. The project was rivaled only by the Panama Canal as an achievement of America's engineering might. The Ashokan Reservoir, bigger than the whole Croton system, with an aqueduct more than twice as long as both Croton pipelines laid end to end, through which water flowed by gravity alone, had the potential to deliver 770 million gallons a day. The reservoir's surface was equal to that of Manhattan below 110th Street. Its contents could fill the Hudson River from the Battery at the southern tip of Manhattan to Hastings-on-Hudson in Westchester County. It took three days for a theoretical drop to travel from the Catskills to Staten Island, which received its first Ashokan water in January 1917. In October 1917, New York City held a three-day celebration of its new water supply, during which 15,000 schoolchildren and a thousand women from Hunter College participated in a pageant in Central Park called "The Good Gift of Water."

It was in 1916 that a letter to the *Times* employed for the first time, at least in that paper, an allegation that was to be slung again and again over the course of the century by a parade of lobbyists on behalf of the city's landlords, alarmed at the prospect of metering. Taps and faucets, it predicted, would become a weapon in the hands of tenants, who would open and run them capriciously for days on end to punish landlords for slights real or imagined. Either apartments were to be metered individually—a patent impossibility—or nothing doing. At that time, houses and apartment buildings were taxed on what were known as frontage and fixture rates, respectively based on the width of the edifice on its street side and the number of stoves it contained, with surcharges for additional stories, families, toilets, and baths. As a 1916 editorial noted, the Water Department acknowledged that these rates "have no relation whatever to the amount of water actually used"[24] and usually represented thirty or forty percent of what the rate should have been. The rates did not, for one thing, figure in depth; if charges were proportionate, an apartment house would owe 32 times as much as a private rowhouse, rather than five and a half times. Furthermore, the system was a bureaucratic

Board of Water Supply, Grand Gorge, NY, postcard, circa 1920

miasma. Two private companies serviced 20,000 meters, while meters in neighborhoods all over the city were served by small independents or village franchisees; the city did not have the right to fix minimum charges that applied to all of them. And there were always problems everywhere with the installation and servicing of meters, whether by licensed plumbers or unlicensed contractors. The editorial noted graphic disparities: In Flatbush, which was unmetered, daily consumption stood at 120 gallons per day per person, while in Newtown, metered, the average was 55 gallons. It applauded a new law, signed by the governor, which gave the city the ownership of meters and the right to maintain them.

Water occasioned much bitter feeling in the city. Thefts of the stuff were widespread. In 1921, an inspector in the Department of Water Supply, Gas and Electric (the body in charge of water seems to have changed names as often as the Parisian vice squad) was arrested, charged with having bought a new seal—an actual wax seal attached to wires hooked to the meter's lock—bearing the number assigned to a large commercial laundry and swapping them. In 1922, eighteen large laundries in the Bronx were suspected of having figured out how to remove and replace seals undetectably, in the meantime having rolled back the meters. The aldermen proposed a law: "No person shall open, hold open, or fasten open, or damage or tamper with any piping or water fixtures"[25] connected to the New York City supply. But the volume of discourse on water meters began to trickle away in the mid-1920s. Now and then a letter writer would disturb the peace with an argument on behalf of universal metering, such as, in 1933, one who made the point that gas and electricity were not sold by flat rate, to which someone else replied that domestic water was unmetered because of the city's concern for the health and sanitation of its inhabitants, and that water could not be compared to gas or electricity because it flushes sewers, cleanses and scours.[26]

Interior of an 8-foot, 4-inch riveted steel-plate pipe at the Tongore Creek crossing of the Catskill Aqueduct, near the Ashokan Reservoir, 1924

C.A.405. CONT. 209. 8.1.24.

1501

3

Gilboa

Panoramic view
of the Gilboa Dam
and the village of
Gilboa, 1922

The Board of Water Supply had already known for years that the Ashokan Reservoir alone would be insufficient in the long run, given the city's ever-increasing population and industrialization. Engineers were surveying for a proposed reservoir on Rondout Creek in Lackawack as early as July 1907; mention of a reservoir on the Neversink River first appeared in the papers in 1911; in 1912, there was talk of new dams on Schoharie Creek and the Delaware River. These projects would all come to fruition, but completion of the full scheme of linked reservoirs in upstate New York would take more than half a century longer.

The first additional project to be undertaken was the dam on Schoharie Creek, in Schoharie County, due north of the Ashokan Reservoir. This undertaking, approved by the city in June 1916, was smaller in scale than its predecessor but represented a devilishly complex engineering challenge. The Schoharie flowed north from the foot of Indian Head Mountain, at the southwest corner of Greene County (which lies between Schoharie and Ulster), into the Mohawk River. Its course had to be diverted south, into a tunnel running from near Prattsville, in Greene County, down to Allaben, on the Esopus northwest of the Ashokan Reservoir. Eighteen miles long, this would be the longest continuous tunnel in the world, its course blasted almost entirely through solid rock. The dynamiting and boring was to be accomplished via seven shafts ranging from 320 to 630 feet in depth; the tunnel's slope would average 4.4 feet per mile.

The reservoir, tapping into a 314-square-mile watershed, had been tentatively planned for Prattsville, which a century earlier had been the national center of the tanning industry until its hemlock forests, required for that process, were decimated. By 1919, engineers had shifted the location to Gilboa, twelve miles north. The dam was to be constructed right atop the site of the village of Gilboa, founded in 1848 through the merger of the villages of Blenheim and Broome, both dating back to the mid-eighteenth century. The village was home to about 350 people; with outlying residents the total number for relocation came to roughly 500. Condemnation proceedings began in 1918, taking 204 properties, with the city operating in its accustomed way—three-man panels, relentless lowballing, dismissals of many claims for damages and loss of business.

The residents, forewarned by the example of their Ulster County cousins, put up fierce if ineffectual resistance, relying on the fact that city employees could not legally enter their houses to turn them out. When people refused to move, however, workers tore off their roofs. Even then some hung on, in empty houses, having prudently removed their furniture. Mary Brooks stood firm until she had to leave for a minute to confer with a neighbor, whereupon workers set her house on fire. A large fire in October 1925, which consumed eighteen buildings, may also have been deliberately set. One man had a brainstorm early on; he built a new house in a critical location, thinking he could soak the assessors. When he was offered a niggling sum, he refused to leave. But he made the mistake of advertising rooms for rent, which gave workers legal right to enter. When the man stepped out on an errand,

Lower Main Street in Gilboa, NY,
postcard, circa 1910

State road crew resurfacing the Grand
Gorge–Prattsville Highway, 1923

The village of Gilboa, 1919

Gilboa-in-the-Catskills Dam, New
Width 1300 feet. Dept

Gilboa Dam, postcard, circa 1926

Form 561E

SCHOHARIE WATERSHED SANITARY RECORD

OWNER ______ NAME Address ______ Date ______

OCCUPANT ______ NAME ______ LOCATION

Number in family ______ Boarders: Regular ______ Summer ______

HOUSE -- Private - Boarding - Hotel - School - Store - Factory - Mill - Slaughter ______

Wooden Frame - Stone - Brick - Stucco ______

Approx. size ______ No. of stories ______ Dist. from watercourse ______

Kind of business ______

BARNS -- Private - Livery - Boarding: Horses ______ Cows ______ Pigs ______

Number ______ Least dist. from watercourse ______

Manure disposed of by ______

OUTBUILDINGS -- Number & condition -- Privy ______ Hog pen ______ Chicken pen ______

Least dist. of each from watercourse ______

DRAINAGE FROM HOUSE: Sink ______ Toilet ______ Bath ______ Cellar ______

Cesspool ______ Dist. from watercourse ______

WATER SUPPLY -- Town, Private ______ If private -- Well, Spring, or ______

DISPOSAL OF GARBAGE AND WASTES ______

REMARKS: ______

Sanitary record card issued to households in the Schoharie watershed, circa 1920

they padlocked the house. *The New York Times* noted in 1925 that of the residents of Gilboa in 1917, "almost a third are dead," and suggested that broken hearts had hastened their end.[1] The 1,330 bodies lying in seven cemeteries also had to be moved. A witness recalled years later that when the graves were opened, the only thing that remained intact were men's silk ties.

A *Times* reporter, writing in 1926 near the end of the project, marveled at the new if temporary city that had grown up in place of Gilboa, with 5,000 residents, paved streets, schools, national and savings banks, electric streetlights, sewers, and 30 miles of railroad tracks with 33 locomotives and 580 cars, as well as 60 derricks and nineteen steam shovels. He may have overestimated the population. The project employed only 1,200 people at any one time, and many of them lived in the individual camps set up at the tunnel shafts, although it is estimated that 10,000 workers were hired overall; grueling conditions ensured a constant turnover. Recruiters for the project were the first in the area to use buses, which shipped in laborers from Philadelphia and Buffalo. The ethnic cross section seems to have been roughly the same as that of the Ashokan workforce: African Americans, Italians, Poles,

Russians, and Swedes, though with a larger representation of Native Americans from northern and western New York.

During tunnel boring in 1921, workers came upon what the press would refer to as a "petrified forest." In the bluestone stratum they found dozens of fossilized stumps, some measuring three feet in diameter, of seed-bearing tree ferns dating to the Devonian era, 300 million years ago. The reservoir, meanwhile, stretched 5.8 miles long, from Gilboa to the outskirts of Prattsville. It was seven tenths of a mile wide and drowned a total of 1,200 acres, with a capacity of 20 billion gallons. The dam, of Cyclopean masonry, was 150 feet wide and 155 feet high. The construction process was awe-inspiring in its scale, with conveyor belts rigged to scaffolding bringing in material from three directions: stone from a nearby

Fossilized tree stumps unearthed during construction of Shandaken Tunnel, 1921

mountain, Portland cement from across the valley, and sand from Prattsville. The reservoir began to be filled in July 1926 during a drought; a massive storm on November 16 brought in eleven billion gallons the first day and a billion gallons each of the next six. An initial dispatch of 660 million gallons headed down the tunnel on November 24, destined for New York City with its 412,000 commercial and residential accounts and 46,000 fire hydrants, where water consumed on a daily basis weighed eight times as much as its residents. The supply at that point was estimated to be ample until 1935. Nevertheless, the city again refused to pay its taxes, worth $60,000 by the time the Schoharie County treasurer threatened to sell the dam a month before the November 16 storm.

380-million-year-old tree fossils from Earth's oldest forest uncovered when stone was being quarried for the Gilboa Dam and currently displayed in locations such as the Gilboa Museum and New York State Museum, September 1, 1922

Welding cradle in Grahamsville, NY, holding three 40-foot lengths of steel pipe conduit destined for the Neversink Tunnel, September 10, 1952

4

Rondout and Neversink

East Delaware Release
Chamber, with Pepacton
Reservoir beyond, 1957

Although surveyed and mapped by 1910, the beginnings of another New York reservoir system, this time on the Delaware River—drawing on a massive watershed that sprawls across large portions of three states—were not undertaken until after the Depression, delayed by lack of funds and by political difficulties as well as by the construction of City Water Tunnel No. 2, crossing the Bronx and the East River into Brooklyn, which usurped the title of world's longest continuous tunnel. (City Water Tunnel No. 1 and City Water Tunnel No. 2 both carry water from the Hillview Reservoir in the Croton system.) The Rondout Reservoir, the first in the Delaware watershed chain, began construction in 1937 but was interrupted twice—by World War II and the Korean War—and was not completed until 1954. Its neighbor, the Neversink Reservoir, 22 miles away, was begun in 1941, just in time for the war, and finished in 1953. Both sit in a valley at the southern end of the

Catskills. The Rondout lies across the border between Wawarsing, at the southwestern end of Ulster County, and Neversink at the northeastern end of Sullivan County. The Neversink is entirely contained within the town of that name. The Rondout collects from Rondout Creek, which rises on Rocky Mountain in the eastern Catskills and describes a counterclockwise course down into Sullivan County before flowing northeast to enter the Hudson at Kingston. The Neversink draws on the Neversink River, a tributary of the Delaware. Each facility was intended primarily to serve as a transfer and collection point from hypothetical future reservoirs farther northwest.

Neversink, NY, circa 1940

In 1937, the hamlets of Eureka, Montela, and Lackawack were condemned in the usual way, consuming 158 parcels in 6,500 acres and displacing around 1,200 residents as well as the remains from eight cemeteries. An unknown number of buildings were destroyed—one source claims a thousand, which seems unlikely—along with unspecified other properties. The largest industry in Lackawack was a tannery; in Eureka it was a grist mill. Perhaps the largest compensation went to one Nora Plunkett of Lackawack, who was awarded $79,000 in 1939, although the specifics of her property do not appear. The valley, not on a railroad line, was relatively remote—the nearest town of any size was Ellenville, population 4,000 in 1940—and events there were not closely covered by the press, which perhaps was bored by then with the water drama. When the Neversink Reservoir displaced 250 people from the hamlets of Neversink and Bittersweet four years later, scarcely a detail of the operation

The old sawmill, Montela, NY, postcard, circa 1910

CANDY
CIGARS
CIGARETTES
Coca-Cola
FOUNTAIN SERVICE
CIGARS, CIGARETTES, CANDY
DRINK
Coca-Cola
BOB PUE
FOUNTAIN SERVICE
CANDY SMOKES
2667-D

The Old Saw Mill, MONTELA, N.Y.

LOOKING DOWN THE NEVERSINK VAL
HUGUENOT, N.Y.
A.11628
ARTINO SERIES BY
C.J. VAN INWEGEN

Looking down the Neversink Valley from
Huguenot, NY, postcard, circa 1910

made the papers. Lackawack and Neversink both relocated, the latter now the hub of its township. The land where the five hamlets had been was to lie stripped and bare for many years.

Besides the Depression, two wars, and the city tunnel, the other matter that delayed construction was a delicate negotiation between the city and the states of Pennsylvania and New Jersey over rights to the Delaware watershed. This matter proved so difficult and time-consuming that the city attempted to undo the Smith Dutchess County Act of 1906, which had barred New York City from taking the waters of Kinderhook, Claverack, and Taghkanic Creeks, Roeliff Jansen Kill, and tributaries of Wappinger, Jackson, Sprout, and Fishkill Creeks, all east of the Hudson—which in the eighteenth century was known as the "Tory" side, as distinct from the west or "Indian" side of the river. That gambit failed, but in 1925 the states managed to hammer out a treaty allocating Philadelphia three billion gallons a day and New York City and northern New Jersey a billion and a half apiece. The two reservoirs and their aqueducts were then approved by the city Board of Estimate in 1928 and by the state the following year. At that point, New Jersey and Pennsylvania brought new objections on technical grounds, but the United States Supreme Court ruled in the city's favor in 1931, although allowing it to take only 440 million gallons a day, a reduction of more than two thirds. Mayor Fiorello LaGuardia, famously pugnacious, feuded with the Board of Water Supply, which he tried to abolish; the state legislature ruled otherwise.

Mayor Fiorello LaGuardia (third from left) and other officials opening the tunnel sluice gates of the Delaware Aqueduct, 1944

The work proceeded quickly, but only when it did so at all. The 85-mile-long tunnel from the Rondout Reservoir to the Hillview, in Yonkers, was excavated at a rate of 137 to 270 feet a week—the Prattsville tunnel had averaged only 55 to 70 weekly feet. The dams were simple earth fills—"unimpressive," according to one author. Yet delays were frequent, in part because wartime rationing made valves and steel unobtainable. On April 5, 1944, when costs had already risen to more than $300 million, LaGuardia triumphantly announced that residents of New York City will drink water from the Delaware water project for the first time tomorrow[1]; a diversion tunnel from

"Stop That Leak!" New York City subway poster by Fred Cooper, circa 1940

Lackawack had been turned on. "The dam can be completed only when we lick Hitler," the mayor added. But even after the war, supply shortages impeded progress, while severe water deficits in 1949 and 1950—which saw the Croton-Catskill system down to 70 percent capacity—made the matter especially urgent.

The next time the subject of water meters arose was, naturally, during a drought. In 1949, as the supply in the reservoirs dropped to 34 percent of capacity, the city suggested that 150,000 dwellings would soon be metered. Although landlords continued to warn against unscrupulous tenants, the Real Estate Board backed the use of meters, proposing that new buildings should be first in line, followed by the buildings with the most inhabitants; landlords should be able to raise rents accordingly. But after the rains came and raised the reservoir levels, the project evaporated.

In November 1952, the 50-billion-gallon Rondout Reservoir was finally completed (although work on the Merriman Dam was not finished until 1954), with a capacity of around 250 million gallons a day, while the 38-billion-gallon Neversink Reservoir was 90 percent done and the Downsville Dam, the first increment in the construction of the Pepacton Reservoir, was 85 percent finished. Authorities anticipated a full yield by 1956: 100 million gallons a day from the Rondout, 105 million from the Neversink, and 335 million from the Pepacton. That same year, the city applied to build a reservoir at Cannonsville on the West Branch of the Delaware, the last in the upstate chain.

The city expected to draw an eventual 900 million gallons a day from the Delaware system, and it hopefully considered that might finally be enough.

The city had been mistrusted and feared upstate before construction of the Catskill reservoirs, but its attitude throughout the sixty-year process—imperious, exploitative, cold—further cemented that mistrust and fear. In 1955, a resident of Cannonsville wrote to the *Binghamton Press* concerning the city's accusation that a local lawyer had created a hostile climate: "I would like to say that Mr. Gottfried has not built the 'hate' for New York City. They have done the deed themselves, but typical of the metropolis they are passing the buck."[2]

In the 1990s, I lived not far from the Pepacton Reservoir, the existence of which I barely knew before I got there, despite having spent the previous twenty-five years as its beneficiary in New York City. Within a month or so of taking up residence in Delaware County, I was struck by the local attitude toward the reservoir. Its construction was spoken of as if it were a disaster—a volcanic eruption, say—that might have occurred decades previously but whose consequences resonated into the present day. Displaced citizens of the drowned villages, who were then still numerous, never felt fully accepted or at home in their new towns; they held regular reunions, with testimonials and home movies, and wrote wrenching reminiscences in local newspapers and self-published books. To be sure, as a former Tompkins town supervisor said of those displaced by the Cannonsville Reservoir—which was completed in 1965—"There were people who couldn't wait to get out, people who took it lying down, and people who took what

C Line
7
4
3
6
8

The Neversink diversion tunnel being constructed through the "drill and blast" method, with white lines superimposed on the image to show desired points of breakage when dynamite was exploded after being inserted into drill holes, February 27, 1942

they could get."[3] But an air of permanent mourning hovered over the county.

For the past twenty years, I've been living close to the Ashokan Reservoir, where the upheaval of construction occurred more than a century ago. Immediate passions may have died out with the generation that experienced the building of Ashokan, but the city is still regarded as an occupying power—like the United States military in Japan, say—that profits from the region while offering little in return and definitely not keeping the best interests of the locality in mind. Residents are subject to strict regulations concerning where they can build and dig and keep livestock; the Esopus Creek below the reservoir trickles along sluggishly until water is released from the spillway, and when released it floods, sometimes ruinously. The taint of the city has spread, as well, via the county's many new residents who come from there, bringing with them liberal attitudes that are regarded with disdain if not outright hostility. The polarization between city and country is an old story, but now it is entrenched in upstate communities.

Measuring cross sections for the Rondout Effluent Chamber, where water now enters the Delaware Aqueduct from Rondout Reservoir, November 15, 1941

Sandhogs in the Delaware Aqueduct, 1937

Team digging in a cramped working chamber during the sinking of two exploratory caissons for the Neversink Dam, September 17, 1941

Paving the Neversink Dam with dry rubble, September 10, 1952

Watershed division clambake,
near Rondout Reservoir, 1947

Construction of the
Pepacton Reservoir
and Downsville Dam,
December 14, 1948

5 The Pepacton Reservoir

Stilling Basin of the
Pepacton Reservoir,
1950

The locale for the Pepacton Reservoir, on the East Branch of the Delaware River, had been surveyed as early as 1907; rumors about land seizures had circulated among the local citizenry at least since 1910. But it was November 16, 1948, when Vincent Impellitteri, president of the New York City Council, finally threw a switch that set off 2,000 pounds of dynamite, creating a 30-by-50-foot hole that represented the first excavation for a dam in Downsville, at the western end of what would be the Pepacton Reservoir. In his speech he said, "It takes a lot of public spiritedness to forgive our tearing down the old homestead where the children were born and in many cases the grandparents and great-grandparents were born and raised." He thanked Delaware County for its "forbearance and friendship."[1]

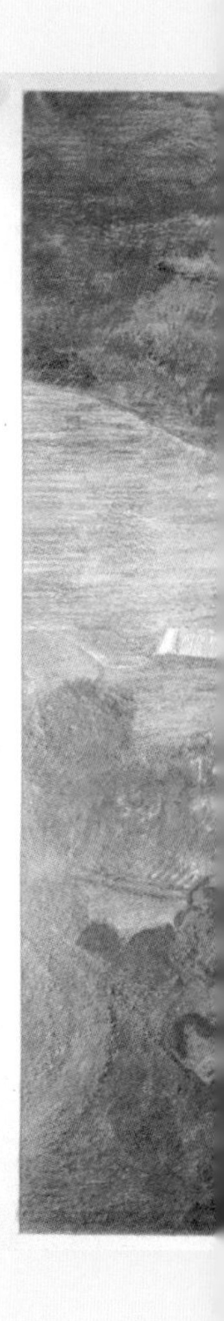

Delaware County, densely forested on its eastern and southern sides, with rolling farmland farther north, looked like an ideal of the northeastern landscape, like a chromolithograph on a calendar issued by a seed company. The great naturalist John Burroughs (1837–1921) was born there, in Roxbury, and in 1881 wrote a book called *Pepacton*, after a then little-used Indian name for the East Branch of the river, said to mean "marriage of the waters."

> The Pepacton rises in a deep cleft or gorge in the mountains, the scenery of which is of the wildest and ruggedest character. For a mile or more there is barely room for the road and the creek at the bottom of the chasm. On either hand the mountains, interrupted by shelving, overhanging precipices, rise abruptly to a great height. . . .
>
> Yet the Pepacton is a placid current. . . . It drains a high pastoral country lifted into long, round-backed hills and rugged, wooded ranges by the subsiding impulse of the Catskill range of mountains, and famous for its superior dairy and other farm products.[2]

Even so, the land was poor. The title of a history of the county conveys the essence: *Two Stones for Every Dirt* (1987).[3] Despite its lush appearance, Delaware County farmland was really only good for a few things: dairy, sheep, and cauliflower, which benefits from a cool climate and does well in the county's abbreviated growing season. By the early twentieth century, many of the mountainsides had been shaved clean of trees, and livestock grazed on sometimes alarmingly canted slopes. The timber, which was lashed into rafts and floated downriver to market, was exhausted by 1898. Many of the people who came to the county in the nineteenth and early twentieth centuries—often from New England—did not tarry long before heading farther west. In 1875, there were 5,449 farms, averaging fewer than ten cows apiece; by 1930, the number had

dropped to 2,913, with an average of 26 cows; by 1969, there were 1,456 smallholdings. Today there are 135 dairy farms left (although beef cattle, not much raised in former times, have taken up some of the slack). Over the decades, suicides of dairy farmers became a serious social issue; in the twenty-first century, the county's suicide rate remains double the state average.

Prospective layout drawing of the Pepacton Reservoir and Downsville Dam showing the reservoir, dam, highway system, and nearby landscaping, July 1943

Despite its strategic importance for the watershed, Delaware County was remote. Legend

Pepacton Reservoir diversion tunnel
drill crew, September 29, 1948

has it that the local Native Americans—primarily Mohawk—did not live there, but used it as a hunting preserve; white settlers did not arrive until the very end of the eighteenth century. It took until 1906 for the first railroad to be built in the Pepacton Valley—the Delaware and Eastern, or, as of 1911, the Delaware and Northern—which connected to the New York, Ontario, and Western at East Branch and to the Ulster and Delaware at Arkville. Its line was sufficiently wild in appearance to pass for Alaska in a 1921 Vitagraph feature, *The Single Track*, directed by Webster Campbell and starring Corinne Griffith and Richard Travers. (It is now lost.) The railway was so underused that in 1926, steam passenger trains were discontinued in favor of a gasoline-powered Brill car—essentially a rail-mounted bus. (It was affectionately called the Red Heifer because its whistle sounded like a moo.) Not many residents owned motor vehicles. In 1950, half the farms had no tractor or truck, a third had no telephone, a tenth had no electricity—and the ones that did mostly obtained it in the wake of the Rural Electrification Act of 1935.

After the initial surveys (and rumors) around 1907, formal surveys for what would become the Pepacton Reservoir were carried out between 1918 and 1923. Exploratory boring was done at Shinhopple in 1927, and the location for a dam at Downsville was decided on in 1933 (reported in the *Downsville News* on April 6 that year). The reservoir's location was chosen in 1938 (reported in the *Catskill Mountain News*, out of Margaretville, on March 18).* Thus it was that nearly a dozen years elapsed during which citizens of Pepacton, Shavertown,

* The *Catskill Mountain News*, founded in 1904, ceased publication in January 2020. There remain a number of small local newspapers in Delaware County—the *Walton Reporter*, the *Deposit Courier*, the *Mountain Eagle* (now merged with the *Schoharie News*)—and they are useful for publicizing 4-H Club benefits, free health screenings, and five-kilometer walks to raise consciousness about rheumatoid arthritis, not to mention providing summaries of sometimes contentious school board meetings and the like. They have nevertheless lost the flavor they possessed back before television and the interstate highway system, when county residents were more connected by family ties and shared circumstances, car accidents were significant news items (and there were an awful lot of them in the 1950s, before power brakes and DUI laws), and visits to or from relatives outside the area were still thought worthy of being noted in print. Their coverage of the reservoir construction and its consequences was often granular and very personal. After the villages of Shavertown and Arena were condemned for the Pepacton Reservoir, the *Catskill Mountain News* published full-page grids of photographs of houses marked for destruction—more than twenty at a time—in full expectation that these pictures would awaken specific emotional associations in their readers.

Union Grove, and Arena were aware of their valley's imminent doom before condemnation proceedings actually began.

The reservoir was to be fifteen miles long by a seventh of a mile across at its widest point, and up to 160 feet deep, draining a watershed of 370 square miles. It remains the largest reservoir in the system, equal to the Ashokan and Gilboa combined and 40 percent bigger than all twelve in the Croton system, holding 140.2 billion gallons at full capacity and supplying a quarter of the city's drinking water. At its eastern end it sends its water through the Rondout Reservoir into the Delaware Aqueduct, which conveys it to the West Branch and Kensico Reservoirs in Westchester County. From there the water flows to New York City, which when Impellitteri threw the ceremonial switch was consuming more than 1.2 billion gallons a day—and there was a drought on; one day in the summer of 1949, the total rose to 1.5 billion. All this infrastructure came at a price: $9 million for land acquisition; $33 million for the Downsville Dam (which was nothing fancy—entirely earthen except for stone facing on its upstream side); $50 million for the tunnel; $5 million for cleaning, grubbing, and grading the land; $13.5 million for replacing the highways; and $2.5 million for replacing the Shavertown Bridge (the single river crossing between Downsville and Margaretville, some 25 miles apart, other than a couple of rickety covered bridges).

The reservoir displaced 974 people: roughly 250 from Shavertown, 175 from Arena, 100 from Union Grove, 70 from Pepacton, and the remainder from outlying farms. In addition, 2,588 bodies had to be moved from thirteen cemeteries, many of them to the new Pepacton Cemetery created for that purpose. The villages were tiny, and of an earlier age. Pepacton, which had one general store and about twenty-five houses, had only been incorporated since 1902, though it had been a center of commercial shad fishing since 1785. Arena, known as Lumbertown until around 1890, began as a single store in 1795. Shavertown, the hub of the valley, was large enough to possess three general stores, six gas stations, three churches, and a barbershop. By 1950, the villages seemed to embody a bygone halcyon era, a vision of "simpler times."

Westerly portion of Cannonsville Cemetery, October 29, 1962

They certainly cast a powerful nostalgic spell on the big-city papers. *The New York Times* and the *Daily News* both sent reporters there for a week or so—three years apart—and both drew heavily on interviews with Inez Atkins, postmaster of Shavertown and owner of one of its general stores, as well as with her sprightly 81-year-old neighbor Amanda Fletcher, who had held the post office job before her. There was also Earl Van Keuren, proprietor of another general store, who was certain he would die when the village did and sat listlessly on the porch while his daughter sold off the remaining stock. (In fact he survived.) And there was Mary Fletcher Thain, beloved of all and an army nurse in World War II, who upon returning bought her riverside dream house (with a window through which she could throw bread crumbs to the fish down below) for only $2,500 because her fellow citizens gallantly

REMOVAL OF BODIES

PLEASE TAKE NOTICE that pursuant to formal notice of the Corporation Counsel of The City of New York the undersigned intends to remove all the remains, all the monuments and all the distinguishing marks from the cemeteries or burial places located in the Town of Neversink, County of Sullivan, New York, known as the Sheley Burying Ground, the Porter Burying Ground, the County Line Farm Cemetery, the Harrison Ryan Farm Cemetery, the George Dierfelter Farm Cemetery and the John Amthor Farm Cemetery.

Please take further notice that any and all persons legally entitled to direct as to the disposition of the above may remove the same to any other cemetery or burial place within the same or an adjoining county within sixty days after September 22, 1939, if they so elect, upon informing James A. Guttridge, Department Engineer, at the office of the Board of Water Supply, Church and Warren streets, Ellenville, New York, in writing, of the cemetery or burial place in which said remains are to be reinterred and to which the monuments and other distinguishing marks are to be moved.

Upon the satisfactory removal of said remains and the said monuments and other distinguishing marks, the Board of Water Supply proposes to make the following allowances to be paid to the person or persons legally entitled to direct as to the disposition of the above upon the presentation to the Board of Water Supply of proper vouchers for payment:

For the removal of the remains, refilling the grave, purchase of new lot and reinterment in a new grave	$40.00 per grave
For the removal and resetting of an ordinary headstone or footstone	4.00 per stone
For the removal and resetting of a monument containing more than one-quarter of a cubic yard	20.00 per cubic yard
For existing foundations under headstone, monument, etc.	0.75 per cubic foot
For removal of fences, copings and steps	0.50 per linear foot

For the removal of anything within the cemetery plot not covered herein or for a removal to a cemetery or burial place outside the same or an adjoining county, a special agreement may be made with the Board of Water Supply upon written application.

Form of notification of removal and reinterment by persons legally entitled to direct the disposition of remains, monuments, etc., and information regarding same, will be furnished upon request at the Ellenville office of the Board of Water Su[illegible]

GEORGE J. GILLESPIE, President,
HENRY HESTERBERG,
RUFUS E. McGAHEN,
Commi[illegible] Board of [illegible]ly of the City [illegible]ork

S. A. HEALY CO. 1806 GENERAL CONTRACTORS

RICHARD H. BURKE, Jr.,
Secretary.

Dated, 348 Broadway, New York, September 8, 1939.

Poster ordering removal of bodies from cemeteries in the town of Neversink, 1939

Margaretville, N. Y., Friday, January 8, 1954 CATSKILL MOUNTAIN NEWS Page Five

A Pictorial Souvenir of Shavertown's Once Happy Homes

Once a part of a living, prosperous community these buildings in Shavertown have been vacated and are being torn down or are about to be torn down as the grubbing and clearing for the Pepacton reservoir bed proceeds.

The News is making an effort to preserve the community in photographs. More of these will be printed in succeeding weeks until the village has been pictured.

THE ARMSTRONG STORE AND FEED MILL was one of the largest structures in Shavertown. Mrs. Hattie Armstrong received $48,900 for this and her home. Bruce Armstrong, who occupied upstairs, has moved to Margaretville.

THE URIAH SPRAGUE store, an old Shavertown business. The case has not been settled. Mr. Sprague has moved to Bloomville.

THE SHAVERTOWN HOTEL and boarding house was owned and operated by Mrs. Carrie Elwood, who was awarded $18,250 by the commission. Mrs. Elwood has moved to Susquehanna, Pa.

MILFORD BUTLER home was originally owned by George Fletcher. The Butlers, who received $10,000 for the house and barn, have moved on to Beech hill.

GEORGE HOAG received $6,800 for this residence alongside the postoffice. Home was formerly owned by Mrs. D. C. Hoag, now deceased.

DON MATTHEWS' garage on the right side of Shavertown's main street drew a commission award of $4,500. Originally the blacksmith shop of Martin Knapp, the building housed an upstairs apartment, which was occupied by Jack Laidlaw.

DON MATHEWS' RESIDENCE alongside the garage also drew an award of $4,500.

HATTIE ARMSTRONG RESIDENCE opposite the Armstrong feed mill was included in the award made for the business property. Mrs. Armstrong is making her residence in Walton.

ED GRASSO residence was formerly owned by Bassett Terry and the late Mrs. Peter Shaver. Grassos have moved to Walton. They received $8,500 in a direct settlement with New York city. House has been occupied recently by Irving Guttridge of the Bureau of Water Supply.

GEORGE REINHARDT residence was awarded $14,000 by the commission for section 17. Mr. Reinhardt is deceased.

LLOYD TERRY RESIDENCE, facing the river, was awarded $13,500 by the commission, but the award was reduced in Supreme court to $10,000 after an appeal by New York city. The Terrys have relocated in the Tremperskill valley.

EVA VAN DUSEN HOME in Shavertown was awarded $7,000. Mrs. Van Dusen and her husband, who also owned a farm on the Union Grove road, have moved to Walton.

PRESBYTERIAN CHURCH PARSONAGE was included in the $34,500 award the church received for its property. House was not used as a parsonage for several years and was occupied by Charles Stanbridge.

ORDWAY ATKIN RESIDENCE. Mr. Atkin, a lifelong resident of Shavertown, has moved to the Elm Tree house in Margaretville. His award for the property in Shavertown was $12,500.

WOODROW LAKIN RESIDENCE was awarded $8,800 by the commission for Shavertown. Lakins have moved to Cannonsville, where New York city is planning its next water supply reservoir.

KARL McCUNE RESIDENCE was awarded $13,500 by the commission. It was formerly the Archie Gladstone residence and faces the river along one of Shavertown's pleasant streets.

WOODROW DENT RESIDENCE was located behind the Atkin store and postoffice. Dents have moved to Walton. They received $8,700 for this property.

FLETCHER HALL was once a creamery and one of Shavertown's principal industries. Hall was scene of community's farewell party on New Year's eve, when Shavertown went out of existence by order of New York city. Mrs. Amanda Fletcher, who owned the hall, received $9,000 for it.

SALVATORE PLATANIA BARBERSHOP AND RESIDENCE occupied this building at the foot of the main street. The Platanias have moved to West Point. The commission valued this building at $9,000.

MARY FLETCHER THAIN RESIDENCE alongside the river drew an award of $10,000. Mrs. Thain is staying at Delhi.

SHAVERTOWN SCHOOLHOUSE has not been used for education since the district was incorporated with the Andes central school district. At one time it housed 125 pupils. There has been no settlement for this property.

ED RUSSELL RESIDENCE was located on the south side of the river by the Shavertown bridge. The Russells are deceased. Settlement for $6,000 was made directly with New York city by the Delaware County Welfare department.

Page from the *Catskill Mountain News* listing Shavertown buildings that would be torn down to construct the Pepacton Reservoir, January 8, 1954

refrained from bidding on it. Now she had done the unthinkable and married a construction worker on the dam from somewhere out of town ("like a Southern belle falling for one of Sherman's soldiers"[4]), and yet she was still beloved.

The *Daily News*, which in 1952 ran a five-part series entitled "This Valley Is *DOOMED!*"—with the final word in a slashing pulp typeface—spent considerable time conducting an inventory of Mrs. Atkins's back shelves: bootjacks, calf-weaners, bulls' nose rings, brass knobs for their horns, Dr. Williams's Pills for Pale People, and Hopping's Driveaway, a panacea for colic, cholera, hysteria, scarlet fever, diphtheria, and "augmented throbbing." There was a large room above the store where square dances occurred, although fewer of them recently. Three years earlier, in the *Times*, Mrs. Atkins had taken an aggressive stance. She charged that "the city has no sentiment," and referred to the authorities' rumored and publicized intentions as "the pestilence that has come down through the valley," pointing out that the 1927 master plan by the Board of Water Supply, which first made known the city's interest in the East Branch, had kept property values depressed for twenty years.[5]

The *Times* tried to put the best face on things: "Farmers, dairymen and storekeepers in the valley's tiny hamlets are in the main alert to the city's needs and recognize the inevitability of the project, even in view of the fact that it means ouster from homes they have built and occupied for generations."[6] Its reporter pointed to Fred Miller of Downsville, who requested $3,000 for his property and received a respectable $2,800, and John Bouw of Holliday Brook, who sold his 600-acre farm for $30,000 and two years later was still living and farming there rent- and tax-free while waiting for the construction crews.* "You can't stand in the way of progress," said a man interviewed by an NBC News crew. But the locals were not all so pliant. In 1950, the county-based *Walton Reporter* published an editorial arguing that New York City should install water meters for its residential customers, which would reduce consumption by 20 percent—reasoning

* These crews may not have arrived; in the end, Holliday Brook was not submerged.

also advanced by the state of New Jersey in its lawsuit twenty years earlier—and that the city should be tapping the Hudson instead. A letter to the *Daily News* by a reader from Narrowsburg praised the paper's series but added: "The only sour note was the statement that all this was designed to provide more drinking water for you N. Y. C. people—at least 30 extra gallons for every man, woman and child. Even a camel couldn't drink that much."[7]

A few months later, the *Catskill Mountain News* reported that lawyers for the city were delaying proceedings "so that they could compile evidence substantiating their contention that farming is not a business."[8] This matter appeared nowhere else in the press. But the city in any case was operating the way it had since the Ashokan seizures, employing condemnation panels of three men (one local, one from the city, and one from elsewhere in the state) and—despite what Mr. Miller in Downsville had received—typically offering about a third of estimates. George Hoag, for example, owned 441 acres with 2,442 feet of river frontage, on which he had a twelve-room house, cow and hay barns, six henhouses, a toolhouse, a springhouse, a milk house, and a garage. He asked for $65,860; the city offered $21,950; they settled on $48,000. Still, land payments were relatively simple overall. More complex were the claim categories that had been written into law since Ashokan: business damage, loss of wages, and

Downsville Dam weir channel, 1952

indirect damages. The valley residents also had a powerful ally: Herman Gottfried, the lawyer accused of fomenting "hate," who had worked for three years in the Board of Water Supply office in Kingston, switched sides in 1950, and represented hundreds of clients affected by the Delaware County reservoirs, despite an attempt by the city to disqualify him for possessing "inside information."

(That case was dismissed by the state supreme court in 1957.)

The legal climate had in fact changed greatly since the Ashokan Reservoir was built. There were more upstate lawyers, for one thing, and they were more attentive to individual complaints, even if no one was in a position to challenge the city directly. Now the city had to pay claimants' witnesses and a portion of their lawyers' fees, and had to compensate farmers for five years' profits in addition to their land, buildings, and equipment. And now anyone who had worked for at least six months at a business adversely affected by the reservoir construction was entitled to six months' wages. Ira Terry and Harry Eckert lost their jobs when the Delaware and Northern was eliminated, but got better jobs elsewhere—Eckert became superintendent of Margaretville Hospital—and so the city tried to deny them their six months, taking the case to the state supreme court, where it lost. Then there were claims for indirect damages. The aforementioned John Bouw, who lived three and a half miles from the village of Pepacton, claimed adverse effect because he now had to go to Downsville—five miles farther—for feed, sawdust, and groceries; he was awarded $3,250 for damage to his land value, and another $2,000 for the additional cost of bringing his milk to the creamery. Stanley Sidorowicz, owner of Stan's Tavern in Arena since 1939, was paid $28,700 by the city in 1955; two years later he sought $50,000 more.

Then there was the matter of trout streams. Fishing rights had been a minor bone of contention in the Ashokan case, as the city dithered for years after the reservoir's completion on whether to allow locals to fish in it (they eventually did). Delaware County, prime angling territory, presented other sorts of problems. In 1948, fishermen worried that the Beaverkill and the Willowemoc, nationally renowned trout streams, would be taken or damaged by the reservoir construction. "It is just as if baseball fans learned that Yankee Stadium and Ebbets Field were to be chopped up into building lots or hot dog stands and shooting galleries," wrote *Daily News* sports columnist Jimmy Powers.[9] (When that didn't happen, local scuttlebutt had it that highly placed anglers in city government had intervened.) In 1957, the Tuscarora Club on Mill Brook, a tributary of the East Branch, sought $125,000

Tuscarora Club Covered Bridge, September 17, 1970

in damages on the grounds that perch and pickerel from the reservoir were invading their trout stream. (The city eventually installed a barrier dam on lower Mill Brook with "iron fingers" to prevent trash fish from entering, while trout used a fish ladder.) The owner of a boardinghouse four miles below the Downsville Dam sought indirect damages because he now had too little water on his property—fishing was spoiled, the swimming hole had shrunk, and the stream was too narrow for boating.

All this was too much for the *Times*, which groused about laws written by "dairy country legislators" to protect their constituents, compensating locals for such things as "esthetic values." "This statutory liberality contrasts with the city's refusal to compensate for business losses when it dislocates its own shopkeepers and business men for slum clearance, road widenings, new schools or other improvements."[10] When the city took property within its own bounds, owners were compensated only for the value of that property, and for fixtures only if they could not be moved. If property was taken for public housing, the city paid tenants $500 for moving expenses; if it

was taken for Title I slum clearance, the shopkeeper could get up to $2,500; if it was taken for a public school or a fire station or for widening streets, the business tenant got zilch. Comparatively, the *Times* grumbled, upstaters were being treated lavishly. The claims in Delaware County did go on and on. In January 1988, there were still 829 claims outstanding from the Pepacton and Cannonsville Reservoirs.

Looking up the river, Shavertown, NY, postcard, circa 1910

But before such claims and counterclaims could begin, of course, the condemnation process itself had to play out, and having begun in 1947, it too seemed to drip on endlessly. Five years elapsed between the first buildings burned in Pepacton and the last ones burned in Arena. Local newspapers marked one final curtain after another, seemingly every week: the auction at the Van Dusen farm near Shavertown, in the family since 1887; the last Christmas pageant at the Union Grove Methodist Church; the last trip on the Delaware and Northern and the last mail at the Shavertown post office, which occurred on the same day; a month later the last mail at the Union Grove post office, which had opened in 1857; the house-by-house sale of every building in Arena, for relocation or scrap. The *Catskill Mountain News* published twelve verses by "a resident of Shavertown":

> Nestled in a peaceful valley
> With the mountains towering high
> Shaded round by shapely maples
> With the river flowing by;
> The iron bridge, the crossroads
> Leading from the mountains down
> This, a meeting place of country folks
> The place called Shavertown.
> [. . .]
> Now the axe of condemnation
> Has fallen on your head,
> Your people all have scattered
> With a listless, weary tread
> To find a place to live in
> Which never will be home,
> Like travelers now in exile
> Cast on the world to roam.
> [. . .]
> Trees uprooted, crushed and broken
> Pushed in piles mountains high,
> And the burnings cast reflections
> On the springtime evening sky.
> Oh Shavertown still hanging
> By a single silver thread
> Which is just about to sever
> And the world pronounce you DEAD.[11]

Tompkins Store, Union Grove, NY,
postcard, circa 1910

Shavertown, postcard, 1906: The
handwritten note says, "The R. R. will be
here in a few weeks."

In October 1955, Mayor Robert Wagner of New York City toured the area, visiting Arena, deserted but still standing, and driving through villages where, as the *Times* crisply observed, "the city's contribution in school taxes sometimes accounts for as much as 90 per cent of the school district's receipts."[12] Wagner was also taken to admire the replacement for the Shavertown Bridge, which had spanned the East Branch. The state supreme court had ordered the bridge to be rebuilt to cross the new reservoir at the same point, but the city had instead elected to erect a 25-foot-high earthen causeway, which cost only $200,000—$2.3 million under the estimate. The causeway had already been battered by Hurricanes Connie and Diane that August; mere hours after the mayor's visit a storm demolished it entirely, washing away its 200-foot midsection, which included the eight tubes that were to have "handled the flow of East Branch water in the upper reservoir."[13] There was no crossing the reservoir at all until August 1956, when a 170-foot steel bridge was completed.

Maple Avenue, Arena, NY, postcard, circa 1910

Three workmen at a pumping station, which keeps water out of the work site of the future Downsville Dam at Pepacton Reservoir, July 12, 1950

17.5-foot-diameter steel form ready to be set during the construction of the diversion conduit for the West Delaware Tunnel, part of the Cannonsville Reservoir system, May 16, 1960

Cannonsville

Groundbreaking at
the West Delaware
Tunnel, 1955

The East Branch is one of several tributaries to the Delaware River proper. It rises near Grand Gorge, in Delaware County, while the West Branch originates near Mount Jefferson, in Schoharie County; they merge near the Delaware County town of Hancock. There is also the Little Delaware, rising near Bovina and joining the West Branch at Delhi. By the mid-1940s, it was generally understood by the authorities concerned that there would need to be a third Delaware County reservoir in addition to the Neversink and Pepacton. A position on the West Branch, near Cannonsville, had been proposed in the 1925 treaty with Pennsylvania and New Jersey, and was finally announced as a firm choice in 1950. Almost immediately, the Delaware County Board of Supervisors passed a resolution against the plan, which would inundate 100 dairy farms that produced more than 2 million gallons of milk per month and remove 800 people from their homes.

The Cannonsville valley is the flattest part of Delaware County and scarcely counts as a portion of the Catskills; topographically, it flows westward into the rolling farmland of the Southern Tier. Once densely forested, it was the last part of the county to receive settlers from points east and south. Jesse Dickinson arrived in 1786 from New Jersey, hacking a trail through the forest, and built a town hall and a grist mill at a place the Mohawks called Ganuissy or Ganiswissa, which he named Dickinson City. But he went broke in the process and sold out to Benjamin Cannon, who renamed the site again. Cannon's house, finished in 1809, was a massive thing with five fireplaces and basement doors big enough for an ox team to drag in six-foot logs. (After his death it served various community functions, then became precinct house for the Board of Water Supply police; it was the last house to be destroyed, in the summer of 1964.) Settlers in significant numbers did not arrive until after 1811, when a road was finally built along the West Branch, from Stamford to Deposit.

When the reservoir's location was decided, the valley contained 74 farms, two creameries, six stores, three post offices, five churches, four schools, two hotels, three restaurants, and a community hall, distributed among the villages of Cannonsville, Granton, Rock Rift, and Rock Royal. There had been a time when the valley bustled, driven by the timber trade and its associated industry, such as factories that produced wood acid; an 1896 Cannonsville directory lists more than a hundred businesses. But all this was contingent on a steady supply of timber, and that gave out in the early twentieth century. The last acid factory closed in 1924, and there remained only Cannonsville's two timeworn sawmills,

descriptively named Slow & Easy and Speedwell. Soon the place was entirely dominated by dairy farming. The pride of the valley was the Rock Royal cooperative creamery, which opened in 1938 and by the end had 153 members, 50 of whom were to lose their land to the reservoir. Besides the creamery, Rock Royal held only a few houses, and much the same was true of Granton, though it did also have

Old Home Day, Cannonsville, NY, 1956

MAP
OF
DELAWARE CO.
NEW YORK
From actual survey
BY JAY GOULD
Published by COLLINS G. KEENEY
PHILADELPHIA
1856
CROTON
DOWNSVILLE
OTSEGO
MEREDITH
DELHI
WALTON
MASONVILLE
HANCOCK
SULLIVAN
STATE OF
PENNSYLVANIA
GRIFFINS CORNERS
MARGARETVILLE
DELHI
J. CHURCHILL
DAGUERREAN ARTIST
DELHI

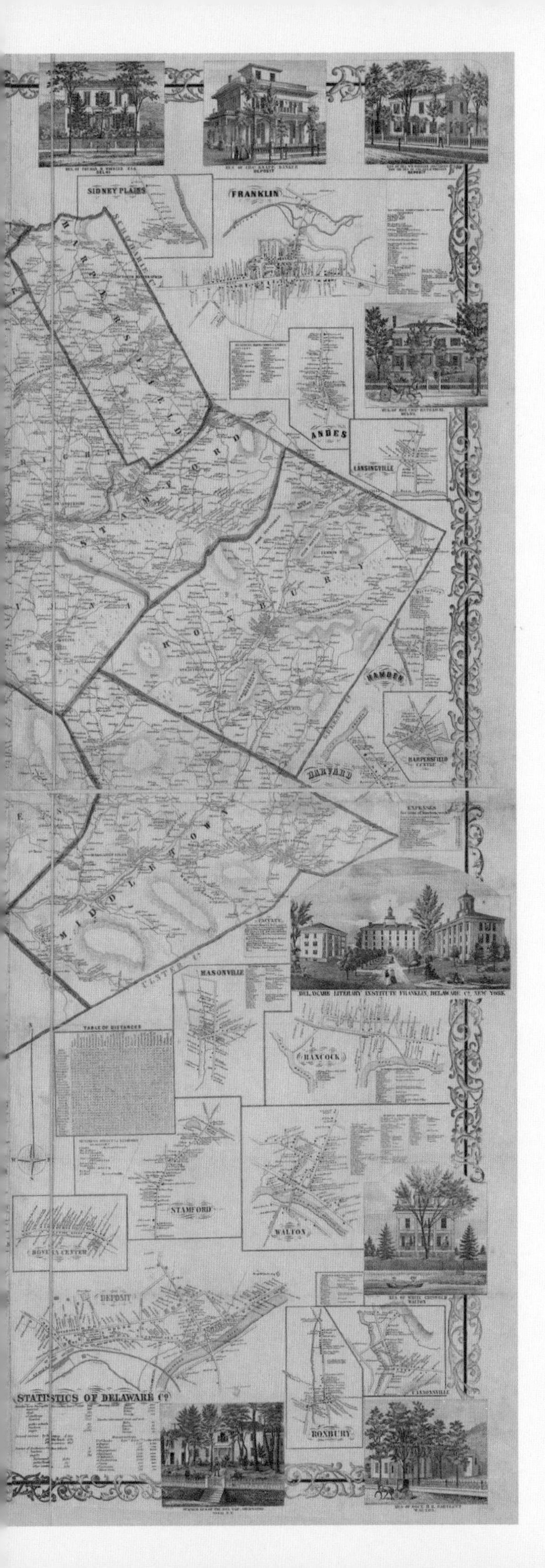

Map of Delaware County, NY, 1856

Rock Royal creamery, circa 1910

a store. Rock Rift had a population of only about 100, many of them lodged in former acid-factory company houses, yet it was a bit busier, with a butcher, a baker, two taverns (Walton, the nearest sizable town, was dry), an Italian restaurant, a general store, and a church. The greater part of the 941 people who would have to relocate lived on outlying farms. The valley also contained eleven cemeteries with 2,150 graves.

The reservoir was to extend sixteen and a half miles from the dam site at Stilesville, near Deposit, to above Rock Rift, with a five-and-a-half-mile finger from Cannonsville up Trout Creek to Rock Royal. The dam was to cut across the western end of a large flat island two and a quarter miles above Stilesville, and to stretch about half a mile wide. The village of Cannonsville was to lie under 75 feet of water.

Route 10 from the dam to Rock Rift would be almost entirely flooded; the dirt road from the dam to Granton would be sunk; the blacktop from Cannonsville to Rock Royal would drown. The maximum width of the reservoir was little more than a mile, at a point opposite Johnny Brook. It would have a capacity of 118 billion gallons and could send up to 323 million gallons a day to the metropolis. The costs were estimated at $20.6 million for the dam and the spillway; $6 million for gatehouse and tunnel construction; a million and a half for clearing and grubbing; $17.7 million for highway relocation, land acquisition, and damages; $980,000 for fencing, planting, and land maintenance; and $7.7 million for administration, engineering, and police, with a 15 percent contingency fund attached, in addition to $80 million earmarked for the tunnel to the Rondout Reservoir. The total $140 million was approved by the New York City Board of Estimate on January 27, 1950. It was expected that the new reservoir would be completed by 1960.

But there was strong opposition to the Cannonsville scheme. At the Board of Estimate hearing, a lawyer for the townships of Walton and Tompkins along with the Rock Royal co-op, William Fancher, argued that farmers would have to ship their milk 40 to 60 miles farther, leading to an increase in prices everywhere. Three months later, at a hearing at Delhi, Lawrence Beck, a consulting engineer appearing on behalf of the valley communities, urged the state Water Power and Control Commission to consider instead a low-level dam across the Hudson near Haverstraw—forty miles north of the city, near the top of the wide part in the Hudson called the Tappan Zee. This dam, of a type known as barrage or gate-pierced (designed to prevent saltwater from passing through), would cost a mere $200 million. But

Benjamin Nesin, director of laboratories for the Board of Water Supply, called the Hudson dam idea "hazardous" and "unreliable," adding, "The Hudson River is virtually a reservoir of infection."[1] At the time, during a period of extended drought, the city was observing "dry Thursdays," a weekly "water holiday" when citizens were urged to keep consumption to the absolute minimum. Momentum was on the side of the plan already in place.

In November 1950, the Cannonsville Reservoir was approved by the state; all that was pending was the sanction of the United States Supreme Court. Beck's engineering plan, which by then included such refinements as ship locks, fish ladders, and a six-lane portion of the New York State Thruway, was rejected. The state noted that the Hudson dam would require an initial capital outlay of $629 million and an annual operating cost of $6 million—and anyway, Beck was unqualified. The argument was taken up again four years later, this time by Harold Riegelman, a brilliant gadfly in city politics who in 1933 had established the independent Citizens Budget Commission to mediate among the mayor, the governor, the legislature, and the bankers in a successful effort to prevent the city from defaulting on its loans. Two decades later, Riegelman, still representing the Citizens Budget Commission, alleged that the city was fiddling with the figures on the reservoir, and had secretly capitulated to the wishes of their counterparts in Pennsylvania—who insisted that Philadelphia have first dibs on Delaware water and that New York should be ready to release 400 million gallons to its neighbor on demand.

In June 1954, the United States Supreme Court amended its 1931 decision, ruling that New York could take an additional 360 million gallons a day from the Delaware, for a daily total of 800 million gallons. "Mayor Wagner said the augmented system should meet the city's water needs until the end of the century," noted the *Times*.[2] On December 14, an unsigned editorial in the paper of record endorsed the idea of a Hudson River dam. A week later, the city's Finance Committee voted to leave undisturbed the $91.6 million initial allocation for the reservoir. In January, the *Times* published a long letter by Riegelman arguing again for the Hudson River plan, and

New York City Mayor Robert F. Wagner and an engineer operating a crane used for the groundbreaking ceremony of the third stage of the Delaware Aqueduct, part of the Cannonsville project, December 19, 1955

The first bucket of earth removed from the West Delaware Tunnel, December 19, 1955

the following month Pennsylvania's attorney general joined the fray, sternly advising New York that it must yield to its neighbor's prior claim on the Delaware and suggesting that a Hudson dam might be a better idea. The fight never really stopped, even through the years when the reservoir was being built. As late as 1963, the Citizens Budget Commission was still issuing press releases calling the Cannonsville reservoir an "inexcusable blunder."[3]

On December 19, 1955, Mayor Wagner broke ground with a clamshell loader at Shaft 2 of the new Delaware Tunnel, near the hamlet of Harvard. The three-man condemnation panels began their work in 1955, with the aim of claiming 19,910 acres, or 3.1 square miles. Herman Gottfried fought for the hearings to be held in Delaware County, while attorneys for the city objected with their claim that Gottfried had "built hate." After a few select acquisitions, such as the 300-acre farm of George and Maude Conrad across from the Cannonsville creamery, and the adjacent 100-acre farm of the Walker brothers, the panels ceased for years, for reasons the record does not make clear. The *Freeman's Journal* of Cooperstown noted some activity in April 1959, after which the panels did not make the news again until June 1961, when state supreme court justice Daniel McAvoy threw out three awards, calling them "shocking and distorted" in their putative generosity. McAvoy, who nine years earlier had cut fourteen awards in the Pepacton cases, denied awards of $171,775 to Mr. and Mrs. Theron Turner for their 205 acres, $123,000 to Mr. and Mrs. Lawrence K. Turner for their 139 acres, and $101,600 to the father and son John and Donald Garrigan. The judge scornfully dismissed expert testimony as "erudite knowledge on water temperature, fish, and aesthetism" and the "unsupported opinion of a so-called real estate expert to the effect that certain land is influenced in value by some hypothetical use to which it might be put."[4] Awards left unchallenged were more modest, such as $65,000 to Kenneth and Eleanor Wright for their 150 acres.

In January 1960, the last rock was blown for the Delaware Aqueduct, and a week later a grand holing-through dinner was held in the Starlight Room at the Waldorf Astoria, honoring the contractors: Johnson, Drake and Piper; the Tecon Corporation; the Conduit and

The West Delaware Tunnel with its steel interlining, protected by an interior "spider" that preserves its roundness during transportation and installation and is later removed, December 1961

Foundation Corporation; and the Winston Brothers Company, descendants of the general contractor who had built the Ashokan Reservoir half a century before. "Each guest at the dinner is to be mailed a handsome office barometer, duly inscribed," remarked the *Catskill Mountain News* in a tone of studious neutrality.[5] Condemnation posters started going up in the valley in the late fall of 1961; in 1962, 298 people moved out. In late January of that year, eleven Cannonsville residents filed an affidavit protesting the city's order to vacate immediately, when it had only acquired title on January 8—and in addition was asking the plaintiffs to move in the dead of winter. Herman Gottfried called the action "faster than Hitler's blitz,"[6] and the court allowed a respite until May.

The *Times* finally sent a reporter to take the local temperature in October 1963. In Cannonsville, he found the graveyard empty of bodies and the creamery with all its windows broken. Valley residents were "vexed" by the actions of the city. Even though the Board of Water Supply had made a point of buying provisions locally and hiring locals for construction and clerical work, they were still bitter. "The farmers here regard the destruction of these communities as wanton. They consider it symbolic of a New York City egomania—an attitude that the outlying provinces exist solely as adjuncts of the city."[7] A woman who moved her house twelve miles from Cannonsville to Stilesville told him, "When they wipe out a whole community—your friends, neighbors, stores, merchants—there's no amount of money that can replace it."[8] Merchants in Deposit were asking for damages; the Western Auto franchisee (at the time an all-purpose working-class emporium) complained that construction workers, many of whom came from the South and didn't plan to stay, couldn't make up for lost trade because they lived in trailers and didn't buy home furnishings. Owners of downriver properties fretted that the dam had made the river too cold for bass, to which the president of the Board of Water Supply responded that they had no right to complain, since the river now contained trout.

The Board of Water Supply opened its Bureau of Claims in 1954 and had handled 200 properties, including more than 80 farms and

100 homes. The last four residents vacated in 1964; the dam was finished on February 16, 1965; the first water went over the spillway in May 1967. The last meeting of the Commission of Appraisal, which since 1935 had reviewed 6,700 claims for a total of $26,806,168, took place in October 1993. Its last major award, in 1992, was $20,598 to Margaretville Hospital for lost business, since 301 patients had moved out of its receiving area.

In 1965, an editorial in *The New York Times* cited the paper's 1860 editorial urging the implementation of water meters. Nothing had changed, it said, except that Robert Wagner was now mayor instead of Fernando Wood. Once again, a water shortage—the reservoirs stood at 55.7 percent of capacity—had prompted a revival of the idea, which would save 220 million gallons a day. "Will it take another hundred years for the city government to come to its senses on water metering as the fundamental conservation measure?"[9] But there was now resistance to the idea from two fronts: landlords and liberals. The Reverend Doctor M. Moran Weston, rector of St. Philip's Episcopal on West 133rd Street, the first African American church in Harlem, inveighed against meters as "another unfair tax on the poor," since it would cause landlords to increase rents. In the 1965 mayoral race, the only candidates wholeheartedly in favor of metering were the liberal Republican John V. Lindsay, who won, and the Conservative, William F. Buckley Jr. The centrist Democrat who would be mayor a decade later, Abraham Beame, hedged his bets; he would be for it if studies demonstrated that savings would follow. He also considered desalination plants as a plausible alternative. But the liberal Democrat Paul O'Dwyer and the Socialist Workers and Socialist Labor candidates firmly opposed the measure as a tax on the poor.

Around 600,000 of the city's buildings were unmetered. (Such figures varied; that number was published in a story dated August 10, 1965, while another appearing ten days later, also in the *Times*, gave the sums as 300,000 private houses and 140,000 apartment buildings.) The cost of purchasing and installing meters was estimated at $100 million. Five years was estimated as the time required to install all the meters. An unnamed member of a municipal panel that urged metering averred that "if water were a private industry, it would never be run the way we run it. Not using meters is like asking them to do business without a cash register." But Mayor Wagner countered that free and unlimited water is "part of the social philosophy of the people of the city" and "a mark of our social advance."[10]

There was now additional pressure from the federal government, in the person of Stewart Udall, the secretary of the interior, who led a delegation of government water experts to city hall for an "open, and sometimes rather hostile," meeting in the Board of Estimates chamber. Mayor Wagner was resistant, citing higher costs associated with metering; water by then cost commercial clients 20 cents per thousand gallons—it could go up to 25.2 cents. A one-family house, under frontage rates, paid $24 a year, while the cost of installing a meter alone was $40. The feds cited the example of Philadelphia, which went to metering in 1955; since then the city's daily consumption had fallen from 400 million gallons a day to 325 million, while New York City was using a billion gallons every summer day. But, countered the municipals, metering worked in Philadelphia because that city consisted mostly of private houses, whereas New York was full of those unscrupulous tenants. In mid-September, the reservoirs dropped to 37.6 percent of capacity (they were to fall to a record low of 25 percent), and the chairman of a Republican "Truth Squad" alleged that Wagner had suppressed a report, unanimously pro-metering, by a committee he had appointed himself. "Slumlords vigorously opposed to water meters . . . heartily welcomed the support of Mayor Wagner as an influential ally."[11]

During the time of New York's financial instability in the 1970s, much was made of the chaos and incompetence that ruled the Board of Water Supply, whose abolition the *Times* called for in 1974. It had a billing backlog of around $70 million. Over a thousand buildings in Brooklyn and Queens were not entered in the billing records and had paid nothing for twenty years or more. Bills for One Lincoln Plaza, the 671-unit luxury condo high-rise across from Lincoln Center that opened in 1971, were sent instead to Dante Park, which lies between them, where the supervisor turned them over to the Parks Department, where they were shredded. Meter reading was so sporadic that in many buildings meters were allowed to complete one or more revolutions and return to zero between inspections, resulting in bills reflecting as little as one third of actual consumption; the discrepancy citywide might have been as much as $17.5 million. The Board of Water Supply was dissolved in 1978 after a corruption scandal pertaining to the construction of City Water Tunnel No. 3, but in 1980, City Comptroller Harrison Goldin estimated that New York was still losing $8 million a year to faulty meters alone.

In January 1981, another drought emergency—reservoirs were at 34.9 percent of capacity, while consumption reached a high of 1.55 billion gallons a day—once again prompted calls for universal metering as well as devices on pipelines that would restrict flow into buildings. These latter, however, had the disadvantage of reducing water pressure, which might cut off upper floors entirely. Restrictions were passed on central air-conditioning (hour-long blackouts in the morning and afternoon in buildings with sealed windows), car washes, laundries, and restaurants, while hydrants were locked. Henry Stern, of the City Council, introduced legislation to require meters in all multiple dwellings, which he estimated would cut consumption, believed to be 75 percent unmetered, by 20 percent; the Rent Stabilization Association—a landlord lobby despite its name—opposed the measure. A separate bill by Stern calling for meters in new and recently renovated buildings was stuck in committee. By the end of the month reservoir levels had fallen to 28 percent, while consumption had risen to 190 gallons per day per person (twenty years earlier the figure had been 154). Officials addressed the matter by proposing regulations on showerheads and requiring aerators on new faucets. Then the rains came and the matter once again disappeared from view.

Another crisis arrived in the summer of 1985 as the reservoirs dropped quickly to around 30 percent capacity. Mayor Edward Koch, hewing to tradition, commissioned a study on whether to compel the city's 600,000 residential buildings to install meters. Nevertheless, in the face of vigorous opposition from landlord groups, he signed Stern's bill requiring meters in new and recently renovated buildings—the *Times* figured that at that rate it would take a century for meters to be installed in all residential edifices. Over the previous twenty years, consumption had increased by a percentage point even as the city's population dropped by half a million, rising to around 207 gallons per day per capita—of which only about one percent was used for drinking. In November, the emergency ended and Koch lifted restrictions. A *Times* editorial was succinct: "To avoid future emergencies, universal metering would be more helpful than autumn rain. . . . New Yorkers do not long remember nature's vagaries and will soon forget the drought emergency of 1985."[12]

To his credit, Koch did not let the matter drop, but urged metering in all buildings, proposing a ten-year calendar for universal installation. Metering of individual apartments was deemed far too costly, however, and the Rent Stabilization Association fought fiercely against master meters. City officials tried to smooth things over by alleging that single-family residences were the key factor, along with regulations on lawn watering, car washing, and swimming pools. Meanwhile, a report proposed the perennial solution of tapping the Hudson River as an additional source. During earlier droughts the city had resorted to importing as much as 100 million gallons a day from the Chelsea Pumping Station, located in Dutchess County north of the salt line but south of the city of Poughkeepsie, which drew its water from the river. But studies showed that the quality of city water suffered as a consequence, with an increase of 80 percent in the pollutant known as PCBs, or polychlorinated biphenyls. These organic chlorine compounds were once widely used as coolant fluids in electrical equipment, among other things. They are toxic to humans' thyroids and nervous systems, and their toxicity lives on for decades. Hudson Valley authorities also took a dim view, deeming the city not authorized to very much in river-water allocation because so much of its consumption was unmetered.

The following autumn the city announced that in one year it would begin a ten-year metering program, at a cost of $290 million, paid for by a nine percent surcharge on water bills on top of a 9.9 percent increase already planned. The president of the Rent Guidelines Board opposed the measure. In 1987, the metering plan was still about to begin, while consumer rates were set to nearly triple, from an average of $148 per year to almost $400 by 1996. In 1988, six areas were designated as part of a metering pilot program: Douglaston, Queens; New Dorp, Staten Island; Cypress Hills, Brooklyn; Riverdale, the Bronx; and East Harlem, Washington Heights, and Inwood in Manhattan. The executive director of the Master Plumbers Council promptly wrote a letter to the *Times* decrying meters as "a hidden tax" and complaining that they would be installed by unlicensed contractors—from out of town—hired by the Water Board through a furtive amendment of the Administrative Code. And furthermore the water system was so decrepit it still had wooden pipes![13] In May of that year the city unveiled a new water-conservation campaign with the slogan "Don't Drip New York Dry!" There was speculation regarding new reservoirs in the Adirondacks. Although the existing reservoirs were at a comfortable 94.5 percent capacity, water would now be served only on request in restaurants.

In March 1989, the reservoirs held only 55 percent of their capacity—normal for the month would have been 91 percent—so the city mandated low-flow toilets in new buildings and urged consumers to take shorter showers and to flush only when necessary. It also reactivated the Chelsea Pumping Station. The Westchester County executive suggested that New York City's water system should be operated by an outside agency—state, regional, or freestanding. By the summer of 1990, 90,000 meters had been installed in single- and two-family dwellings, with 470,000 to go. Landlords were said to be considering abandoning properties rather than face the costs of metering. In August, under state pressure, the city formally committed to installing 600,000 meters by 1998—its first legal commitment to the project.

In 1991, Vanguard Meter Services of Owensboro, Kentucky, which had won 22 of the 32 metering contracts, began to be investigated by federal and state prosecutors. They were said to underpay employees; there were serious questions about how they came to be the sole

bidder on several contracts and how their only competition in others came from firms that happened to share their Long Island City offices. Six months later the top executives of Vanguard were indicted for fraud. As installation of meters continued by other firms, there were mounting complaints about dramatic rate increases in some buildings, although not in others. The city inaugurated a program in which landlords were compensated for installing water-efficient toilets. Usage measurably dropped in Douglaston and New Dorp.

In the summer of 1999, nearly 65 percent of the roughly 46,000 multifamily dwellings in the city still did not have meters. Owners who had not installed them by June 2000 would face a 100 percent surcharge. At the end of the year, as the deadline loomed, 60,000 buildings were still unmetered. (Notice how fluid all these estimated figures seem to be.) In December 2000, when the deadline had come and gone, the city offered the landlords of 72,000 buildings of six or more units that were still unmetered the option of a flat rate, with the stipulation that those buildings must still have meters installed, although they would be billed at a standard $424 per apartment.

The Jacqueline Kennedy Onassis Reservoir, May 24, 2012

In 2006, when the city's population hit a then-record high, its water use had notably declined to 1.086 billion gallons a day, the lowest since 1951, when there was a drought on. The city used 28 percent less water than it did in 1979. The increased popularity of bottled water and the decline of manufacturing were cited as factors, but the largest was the cumulative effect of all those years of warnings about wasting water and the gradual installation of dripless faucets and low-flow toilets. By the end of that year, however, the city was far behind on collecting its water bills by as much as $630 million. (Then again, one building with an outstanding debt of $171,879 was discovered to have been torn down in 1990.) By then, 96 percent of the 826,000 individual water accounts were metered, but the Water Board's rate schedule still included prices for filling steamboats, sidewalk horse troughs, and milk depots. It was noted that New York City had never once turned off water service to anyone for nonpayment.

In the meantime, troubles continued for the Cannonsville Reservoir, the water of which was generally deemed to be of lesser quality than that of the others in the system. A plan by New Jersey and Pennsylvania to build a new reservoir at Tocks Island, near the Delaware Water Gap, failed in 1976. The following year New Jersey, Pennsylvania, and Delaware threatened to go to the Supreme Court to keep New York State—which had assumed responsibility a year earlier—from controlling the flow on the three Delaware reservoirs. That year the Delaware River Basin Commission unanimously approved a plan requiring New York to release an additional 31 billion gallons a year from Cannonsville, Pepacton, and Neversink. In 1996, a flood in Walton caused the road ringing the Cannonsville Reservoir to collapse, throwing four cars down 70 feet and killing six people. Floods in 2004, 2005, and 2006 killed nine and caused $500 million in property damage in four states. In 2007, after a long dispute, New York finally agreed to release up to 35 million gallons a day back into the Delaware to maintain temperature and water levels. By 2017, city officials worried that rising ocean levels might endanger New York's drinking water, since ever more reservoir water would be needed to maintain the salt line in Delaware Bay. Like the Hudson, the Delaware is tidal; its water is fresh at least down to Trenton, New Jersey, and is considered brackish at Wilmington, Delaware. Regular flushing with fresh water is required to keep the brackishness from creeping northward.

Back in 1997, however, the city's Department of Environmental Protection had been reassuring: "Based on the results of metering, toilet replacement, public information, and other conservation programs achieved to date and expected in the future, it is projected that no additional water sources will be necessary for the next fifty years."[14] On close inspection, this does not truly reassure. But it has been a while since the Northeast has faced a serious drought, and no incidents or complaints have recently made the news, so the health of the system has not been questioned of late, at least not in public. In the meantime, New York City continues to take its drinking water from those artificial lakes in mountain valleys, so inviting on a hot day in a region with no real lakes, albeit as taboo for swimming or boating as if they were meant for the gods alone. The ghosts of the drowned villages continue to haunt the popular imagination via roadside markers and twice-told tales, even as the tally of relocated citizens—never quite integrated into their new communities, in places where "away" might mean five miles and "new" might mean fifty years ago—continues to diminish. Farms continue to fail, and farmers continue to die, and land and houses continue to be bought by city people who wouldn't know their sheep-dip from their cream separator. New York, like other cities, is filled with people who have no idea where their water comes from and are only occasionally made aware that it is a precious and very finite resource that will become scarce again one day—perhaps quite soon. By then there will be no untapped mountain valleys to draw from.

Olive Bridge Dam
on the Ashokan
Reservoir, 2020

Epilogue

It has been more than a century since the Ashokan Reservoir was put into service, and more than fifty years since completion of the Cannonsville Reservoir, the last in the series. All the reservoirs have by now merged into their landscapes as if they had always been there. During the summer of 2020, Tim Davis spent weeks photographing those landscapes from close up, far away, and all points in between. Accordingly, his view of them ranges across the spectrum, from their unwitting role as calendar images of lacustrian perfection to the cultural undergrowth, gnarly or serene, that surrounds them.

Is it because we know the reservoirs are artificial that their very beauty can appear confected? Selfie takers smile performatively at their camera; in the background the Ashokan looks like a matte painting from 1950s cinema. It can seem as if the reservoirs were engaged in supplying not just drinking water to consumers, but also an especially consumable type of destination, a bit hyperreal. But then the water reflects clouds that appear to come from an old-master painting, and the scene is so primordial the dam in the distance just looks like a piece of tape on the canvas.

Meanwhile, the most evident litter continues to be produced by beavers. A tree is gradually swallowing a sign for Hidden Hollow, and maybe the hollow itself was already swallowed up. Bluestone quarries, exhausted long ago for the sidewalks of a dozen cities, can look like so many empty graves. The graveyards themselves, set high in the hills, look as majestic as war memorials, while the highways are dotted with signs—"Former Site of Glenford"—that suggest grudging civic remembrance of something best forgotten.

At the western end of the Ashokan, Snyder's Tavern is a haunted house with neon beer signs in the windows; people have been seen butchering deer in the parking lot. On the eastern side is Kenco, a recreational outfitter that promises a "Safe, Healthy Outdoors" to city people who are wary of nature. Locals fly the collie flag and wave the Trump banner, work on their cars and populate their land with metal sheds, maybe because the wooden ones have fallen down and maybe because they keep accumulating stuff. The villages can sometimes seem unfinished and sometimes abandoned, but a town meeting—held outside due to COVID-19—can appear as intimate and involving as a farm auction.

A mere fraction of farms survive from the time before reservoirs, but the reservoirs themselves are not the culprit. The topography is

Ashokan Reservoir, 2020

too variegated, the plots too small, and the soil too stony for the sort of agriculture that today's markets demand. At this point, country people might just keep a couple of cows for their own use while they go to work repairing roads or installing seamless gutters or tending their circuit as a home health aide—or doing landscaping for the city people who now own their ancestors' land. Anyway, they can always go fishing in their regulation steel rowboats, floating in vast acreages of still water their ancestors never knew.

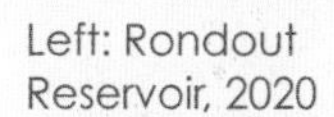

Left: Rondout Reservoir, 2020

Top: Downsville Covered Bridge, 2020

Bottom: Ashokan Reservoir, 2020

Boiceville, NY, 2020

Reservoir Motel in Shokan, NY, 2020

Gilboa, NY, 2020

Angler at Pepacton Reservoir, 2020

All photographs of
Downsville, NY, 2020

DOWNSVILLE
"BEST TOWN BY A DAM SITE"
STORE
MOTEL
DINER
GLADSTONE INSURANCE
Catskill Country Market
CATSKILL CELLARS WINES & SPIRITS
D & N Improvements, LLC
LORRAINE'S HAIR SALON
FLO'S YARN & CRAFT SHOP
Upper Delaware Real Estate
Community Bank N.A.
MATT RAY PLUMBING
ZELLER INTERNATIONAL LTD.
HEAD-TO-TOE FITNESS

Boiceville, NY, 2020

Shokan, NY, 2020

Conesville, NY, 2020

West Shokan, NY, 2020

Town meeting, Grahamsville, NY, 2020

Harvard, NY, 2020

Lower spillway, Olive Bridge Dam, 2020

Grahamsville, NY, 2020

Offices of the *Sullivan County Democrat* newspaper in Callicoon, NY, 2020

Route 28, Ulster County, NY, 2020

Hancock, NY, 2020

Dragging a boat to the Pepacton Reservoir, 2020

Oxbow Campsites, Harvard, NY, 2020

Rondout Reservoir, 2020

Olive Bridge Dam
on the Ashokan
Reservoir, 2020

Notes

THE CROTON SYSTEM

1. Moses King, *King's Handbook of New York City* (Boston, MA: Moses King, 1892), 181.
2. DeWitt Clinton Jr., quoted in Charles H. Weidner, *Water for a City: A History of New York City's Problem from the Beginning to the Delaware River System* (New Brunswick, NJ: Rutgers University Press, 1974), 29.
3. "Cheap Water," *The New York Times*, October 9, 1860.
4. Pro Bono Publico, letter to the editor, *The New York Times*, 1870.
5. "The Tweed Water Meters," *The New York Times*, October 13, 1880.
6. Property Owner, letter to the editor, *The New York Times*, December 24, 1899.

7. W. Volkhardt, letter to the editor, *The New York Times*, February 20, 1902.
8. "Water Saving May Stop Catskill Plan," *The New York Times*, January 16, 1910.
9. "Water Meters in Homes," *The New York Times*, November 8, 1912.

THE ASHOKAN RESERVOIR

1. R. D. A. Parrott, "The Water Supply for New York City," *Scientific American Supplement* no. 697 (May 1889): 11132.
2. Parrott, "The Water Supply for New York City," 11132.
3. Parrott, "The Water Supply for New York City," 11131.
4. Parrott, "The Water Supply for New York City," 11131.
5. "Looking Over Property," *Kingston Daily Freeman*, May 16, 1907.
6. "Shokan Puzzled," *Kingston Daily Freeman*, March 1, 1905.
7. Parrott, "The Water Supply for New York City," 11132.
8. Parrott, "The Water Supply for New York City," 11132.
9. Diane Galusha, *Liquid Assets: A History of New York City's Water System* (Fleischmanns, NY: Purple Mountain Press, 1999), 93.
10. "Passage of Coutant Bill Strongly Urged," *Kingston Daily Freeman*, March 16, 1905.
11. *Kingston Daily Freeman*, May 7, 1908.
12. *Kingston Daily Freeman*, July 12, 1907. Delaney was quoting a popular song of the time: "Little Willie stood a-watching / While his father dug a well; / Little Willie lost his footing— / Cheer up, boys, there ain't no hell."
13. "Damage for Digging Pits," *Kingston Daily Freeman*, June 14, 1907.
14. "Water Board Has Funds to Carry on Its Work," *The New York Times*, September 14, 1906.
15. "Ginseng Garden Up in Olive," *Kingston Daily Freeman*, November 27, 1907.
16. Charles H. Weidner, *Water for a City: A History of New York City's Problem from the Beginning to the Delaware River System* (New Brunswick, NJ: Rutgers University Press, 1974), 232.
17. Weidner, Water for a City, 244–45.
18. "Model Camp Built at Ashokan Dam," *The New York Times*, May 2, 1909.
19. "Aqueduct Cops Let Sobo Go," *Kingston Daily Freeman*, July 16, 1908.
20. "City to Be Built at the Ashokan Dam," *The New York Times*, September 7, 1908.

21. "Burhans Describes Brown's Station," *Kingston Daily Freeman*, July 26, 1907.
22. "Oklahoma and Ashokan," *Kingston Daily Freeman*, August 15, 1910.
23. "Esopus Dam Closed, Water Famine Now Impossible," *The New York Times*, October 12, 1913.
24. "Water Rates and Meters," *The New York Times*, May 15, 1916.
25. "Wasting the City's Supply of Water," *The New York Times*, July 1, 1917.
26. Theodore Reed Kendall, letter to the editor, *The New York Times, September 11, 1933; and John Yearwood, letter to the editor, The New York Times*, September 14, 1933.

GILBOA

1. Alvin Meland, "Gilboa Engulfed by New York," *The New York Times*, September 13, 1925.

RONDOUT AND NEVERSINK

1. "Delaware Water Flows Here Today," *The New York Times*, April 5, 1944.
2. Letter to the editor, *Catskill Mountain News*, November 25, 1955.
3. Diane Galusha, *Liquid Assets: A History of New York City's Water System* (Fleischmanns, NY: Purple Mountain Press, 1999), 224.

THE PEPACTON RESERVOIR

1. "New York Sorry to Despoil Valley Say Speakers," *Catskill Mountain News*, November 19, 1948.
2. John Burroughs, *Pepacton* (Boston: Houghton Mifflin, 1881), iii–iv.
3. Douglas DeNatale, *Two Stones for Every Dirt: The Story of Delaware County, New York* (Fleischmanns, NY: Purple Mountain Press, 1987).
4. Joe Martin and Kermit Jardiker, "This Valley Is Doomed!" *New York Daily News*, October 9, 1952.
5. Kalman Seigel, "Reservoir Evicts Many from Homes," *The New York Times*, June 4, 1949.
6. Seigel, "Reservoir Evicts Many from Homes.".
7. Letter to the editor, *New York Daily News*, October 16, 1952.
8. "Commission Hears Several Cases at Downsville," *Catskill Mountain News*, December 22, 1950.
9. Jimmy Powers, *New York Daily News*, September 8, 1948
10. Charles Grutzner, "Upstate Water Claims Flood City," *The New York Times*, July 28, 1957.

11. "Shavertown," *Catskill Mountain News*, April 9, 1954.
12. Charles, G. Bennett, "City Water Basin Impresses Mayor," *The New York Times*, October 15, 1955.
13. "Causeway Washout Leaves Valley Lacking Bridges Across Reservoir," *Catskill Mountain News*, October 21, 1955.

CANNONSVILLE

1. "Water Use in City Continues to Rise," *The New York Times*, June 8, 1950.
2. "City Gains Water on Delaware Tap," *The New York Times*, June 8, 1954.
3. Budget Group, "Cannonsville Dam Called an 'Inexcusable Blunder," *Catskill Mountain News*, December 26, 1963.
4. "Commission Awards Refuted by Justice," *Catskill Mountain News*, June 16, 1961.
5. "Contractors Gave Dinner to BWS," *Catskill Mountain News*, January 29, 1960.
6. "Cannonsville Residents Go to Court to Stave Eviction," *Catskill Mountain News*, January 25, 1962.
7. "City Panel Backs Water Metering in All Buildings," *The New York Times*, August 20, 1965.
8. Martin Tolchin, "Farmers in Cannonsville Area Vexed by City's New Reservoir," *The New York Times*, October 22, 1963.
9. "As We Were Saying—On Water," *The New York Times*, May 13, 1965.
10. McCandlish Phillips, "City Panel Backs Water Metering in All Buildings," *The New York Times*, August 20, 1965.
11. Peter Kihss, "Westchester Attacks Wagner's Refusal to Order Water Meters," *The New York Times*, September 14, 1965.
12. "The Lesson of November's Rains," *The New York Times*, November 30, 1985.
13. Stanley Gold, letter to the editor, *The New York Times*, February 20, 1988.
14. Galusha, op. cit., 225.

Acknowledgments

I would like to thank, first of all, Frances Richard and Nancy Levinson at *Places Journal*, who took on my wisp of an idea and allowed it to flourish; my wonderful agent, Joy Harris, who thought my article series should be a book; Melissa Holbrook Pierson, for her loan of materials; and Matthew Lore, Hannah Matuszak, Beth Bugler, and the whole team at The Experiment. And special thanks to Mimi Lipson, as ever.

Image Credits

Every effort has been made to trace and contact copyright holders. If an error or omission is brought to our notice, we will be pleased to correct it in future editions of this book. For further information, please contact the publisher.

Pages i, vii, viii–ix, 11, 12–15, 17, 19, 26, 34, 42–43, 44–45, 49, 51 (top and bottom), 52–53, 53, 56–57, 60–61, 65, 66–67 (bottom), 68, 72–73, 79 (bottom), 80–81, 84, 85, 86–87, 88–89, 90–91, 92–93, 96–97, 98, 100–1, 102–3, 104 (top and bottom), 105, 106–7, 108–9, 110–11, 112–13, 114–15, 118, 122–23, 132–33, 134–35, 136–37, 145, 146, 148–49, 174–75, courtesy of the New York City Department of Environmental Protection.

Pages ii–iii, 6–7, 8–9, 10, 24–25, 36–37, 39 (top and bottom), 40, 54, 58, 62, 70, 79 (top), 82–83, 93, 94–95, 119, 120, 126–27, 129 (top and bottom), 130–31, 138–39, 142–43, courtesy of Lucy Sante.

Pages iv–v, Adobe Stock/Eyetronic.

Pages x–1, Currier & Ives. The city of New York, 1892. https://www.loc.gov/item/75694825.

Pages 2–3, Silver, William W. *City and harbor of New York*, 1896. https://www.loc.gov/pictures/item/2007661278.

Pages 4–5, Detroit Publishing Co. *Mulberry St., New York, N.Y.*, 1900. https://www.loc.gov/pictures/item/2016800172.

Page 16, The Miriam and Ira D. Wallach Division of Art, Prints and Photographs: Print Collection, The New York Public Library. "Reservoir of Manhattan water works. Chamber St., 1825." New York Public Library Digital Collections. https://digitalcollections.nypl.org/items/510d47da-f381-a3d9-e040-e00a18064a99.

Page 18, The Miriam and Ira D. Wallach Division of Art, Prints and Photographs: Picture Collection, The New York Public Library. "Croton Dam" New York Public Library Digital Collections. https://digitalcollections.nypl.org/items/510d47e1-06f6-a3d9-e040-e00a18064a99.

Page 21 (top and bottom), 74–77, courtesy of Municipal Archives, City of New York.

Page 23, Nast, Thomas. *Boss Tweed. Published by Harper's Weekly*, 1871.

Pages 28–29 (top), Lionel Pincus and Princess Firyal Map Division, The New York Public Library. "Map and profile showing sources of, and manner of obtaining, an additional supply of water for the city of New York," The New York Public Library Digital Collections. 1905. https://digitalcollections.nypl.org/items/98502d00-757e-0131-09a2-58d385a7bbd0.

Pages 28–29 (bottom), 22. *Photocopied from New York City Aqueduct Commission. 1907 plan of downstream elevation - New Croton Dam & Reservoir, Croton River, Croton-on-Hudson, Westchester County, NY*, 1907. https://www.loc.gov/pictures/item/ny1163.photos.124488p.

Pages 30–33, 66–67 (top and middle), courtesy of Mark Dubois

Page 125, copyright © Wikipedia 2013, "Tuscarora Club Covered Bridge," under Creative Commons Attribution-ShareAlike 3.0 Unported License.

Pages 140–41, Gould, Jay. *Map of Delaware Co., New York*. Published by Collins G. Keeney, 1856. https://www.loc.gov/item/2012593655.

Pages 156–57, copyright © Wikipedia 2018, "Jackie Onassis Kennedy Reservoir," under Creative Commons Attribution-ShareAlike 2.0 Generic License.

Pages 160–61, 162–63, 164, 165 (top and bottom), 166 (top and bottom), 167, 168–69, 170 (top and bottom), 170–71, 171, 172 (top and bottom), 172–73, 173 (all), 174–75, 176 (all), 176–77, 177, 178, 178–79, 179, 180–81, 187, Tim Davis.

Index

NOTE: Page references in *italics* refer to figures and photos. Page references followed by *n* refer to footnotes.

About Lucy Sante and Tim Davis

Lucy Sante was born in Verviers, Belgium, and is the author of eight other books. She is the recipient of a Whiting Award, Guggenheim and Cullman Center Fellowships, an Award in Literature from the American Academy of Arts and Letters, a Grammy (for album notes), and an Infinity Award for Writing from the International Center of Photography. Sante has contributed to *The New York Review of Books* since 1981 and to many other publications. She recently retired after twenty-four years teaching at Bard College and lives in Ulster County, New York.

lucysante.com | @luxxante

Tim Davis, born in Malawi in 1969, is an artist, writer, and musician who lives in Tivoli, New York, and teaches photography at Bard College. His latest book, *I'm Looking Through You,* was published by Aperture in 2021. In 2019, the Tang Museum at Skidmore College showed a large survey of his recent work in photography, video, sound, sculpture, and performance, entitled *When We Are Dancing, I Get Ideas.* His work is in the permanent collections of the Guggenheim, Whitney, Brooklyn, and Metropolitan Museums in New York; the Milwaukee Museum of Art; the High Museum of Art in Atlanta, Georgia; the Baltimore Museum; the Smithsonian's Hirshhorn Museum and Sculpture Garden; the Los Angeles County Museum of Art; and numerous others.

ohthattimdavis.com | @ohthattimdavis